Managing Complex Technical Projects:
A Systems Engineering Approach

For a complete listing of the *Artech House Technology Management and Professional Development Library*, turn to the back of this book.

Managing Complex Technical Projects: A Systems Engineering Approach

R. Ian Faulconbridge
Michael J. Ryan

Artech House
Boston • London
www.artechhouse.com

Library of Congress Cataloging-in-Publication Data
Faulconbridge, Ian.
  Managing complex technical projects: a systems engineering approach / Ian Faulconbridge,
Michael Ryan.
    p. cm. — (Artech House technology management and professional development library)
    Includes bibliographical references and index.
    ISBN 1-58053-378-7 (alk. paper)
      1. Systems engineering.   2. Project management.   I. Ryan, M. J. (Michael J.)
II. Title.   III. Series.
TA168. F29 2002
658.4'04—dc21   2002032660

British Library Cataloguing in Publication Data
Faulconbridge, Ian
  Managing complex technical projects: a systems engineering approach. — (Artech House
technology management and professional development library)
  1. Project management   2. Systems engineering
  I. Title   II. Ryan, Michael
  620'.00685

  ISBN 1-58053-378-7

Cover design by Yekaterina Ratner

© 2003 ARTECH HOUSE, INC.
685 Canton Street
Norwood, MA 02062

International Standard Book Number: 1-58053-378-7
Library of Congress Catalog Card Number: 2002032660

10 9 8 7 6 5 4 3 2 1

Contents

Preface   *xiii*

Introduction to Systems Engineering   **1**

1.1   What Is a System?   2

1.2   System Life Cycle   5
1.2.1   *Acquisition Phase*   6
1.2.2   *Utilization Phase*   8

1.3   What Is Systems Engineering?   9
1.3.1   *Requirements Engineering*   10
1.3.2   *Top-Down Approach*   10
1.3.3   *Focus on Life Cycle*   11
1.3.4   *System Optimization and Balance*   13
1.3.5   *Integration of Disciplines and Specialties*   14
1.3.6   *Management*   14

1.4   Systems Engineering Relevance   15

1.5   Systems Engineering Benefits   16

1.6   Analysis, Synthesis, and Evaluation   18
1.6.1   *Analysis*   19

1.6.2	*Synthesis*	20
1.6.3	*Evaluation*	20
1.7	A Systems Engineering Framework	21
1.7.1	*Systems Engineering Processes*	23
1.7.2	*Systems Engineering Management*	23
1.7.3	*Systems Engineering Tools*	23
1.7.4	*Related Disciplines*	24
	Endnotes	24
2	**Conceptual Design**	**29**
2.1	Introduction	29
2.2	Identify Stakeholder Requirements	30
2.2.1	*Stakeholder-Requirements Document*	30
2.2.2	*Identify Stakeholders*	32
2.2.3	*Identify Project and Enterprise Constraints*	33
2.2.4	*Identify External Constraints*	34
2.2.5	*Define Need, Goals, and Objectives*	34
2.2.6	*Define Operational Scenarios*	35
2.2.7	*Define Measures of Effectiveness*	36
2.2.8	*Define Life-Cycle Concepts*	36
2.2.9	*Confirm SRD Structure*	36
2.2.10	*Scoping the System*	38
2.2.11	*Populate SRD*	40
2.2.12	*SRD Endorsement*	40
2.2.13	*Traceability*	41
2.3	System-Feasibility Analysis	41
2.4	System-Requirements Analysis	43
2.4.1	*Establish Requirements Framework*	45
2.4.2	*Define Functional Requirements*	47
2.4.3	*Define Performance Requirements*	49
2.4.4	*Define Verification Requirements*	50
2.4.5	*Assign Rationale*	50
2.4.6	*Perform Functional Analysis and Allocation*	51

2.4.7	*Produce Draft System Specification*	55
2.4.8	*Define TPMs*	55
2.4.9	*System-Requirements Reviews*	58
2.4.10	*Other System-Level Considerations*	59
2.5	System-Level Synthesis	60
2.6	System-Design Review	63
	Endnotes	65

3	**Preliminary Design**	**67**
3.1	Introduction	67
3.2	Subsystem-Requirements Analysis	68
3.3	Requirements Allocation	72
3.4	RBS Versus WBS	80
3.5	Interface Identification and Design	82
3.6	Subsystem-Level Synthesis and Evaluation	85
3.6.1	*Review Sources of Subsystem Requirements*	85
3.6.2	*Investigate Preliminary Design Alternatives*	86
3.6.3	*Make Optimal Use of Design Space*	88
3.6.4	*Select Preferred Solution*	93
3.7	Preliminary Design Review	94
	Endnotes	96

4	**Detailed Design and Development**	**97**
4.1	Introduction	97
4.2	Detailed Design Requirements	98
4.3	Designing and Integrating System Elements	98
4.3.1	*Detailed Design Process*	99
4.3.2	*Integration*	100
4.3.3	*Some Detailed Design Aids*	103
4.4	System Prototype Development	104

4.5	Detailed Design Reviews	105
4.5.1	*Equipment/Software Design Reviews*	105
4.5.2	*Critical Design Review*	106
4.6	Construction and Production	108
4.7	Operational Use and System Support	112
4.8	Phaseout and Disposal	117
	Endnotes	119
5	**Systems Engineering Management**	**121**
5.1	Introduction	121
5.2	Technical Reviews and Audits	121
5.2.1	*Major Reviews*	123
5.2.2	*Major Audits*	125
5.2.3	*Technical Review and Audit Management*	126
5.3	System Test and Evaluation	127
5.3.1	*Developmental Test and Evaluation*	129
5.3.2	*Acceptance Test and Evaluation*	130
5.3.3	*Operational Test and Evaluation*	131
5.3.4	*Test Management*	132
5.3.5	*Testing Activities and the System Life Cycle*	132
5.3.6	*TEMP*	137
5.4	Technical Risk Management	139
5.4.1	*Risk Identification*	140
5.4.2	*Risk Quantification*	141
5.4.3	*Risk-Response Development and Control*	142
5.4.4	*Risk-Management Documentation*	143
5.5	Configuration Management	144
5.5.1	*Establishing the Baselines*	145
5.5.2	*Configuration-Management Functions*	145
5.5.3	*Configuration-Management Documentation*	151
5.6	Specifications and Standards	152

5.6.1   *Specifications*   153
5.6.2   *Standards*   156

5.7   Integration Management   157

5.8   Systems Engineering Management Planning   159

        Endnotes   160

6   Systems Engineering Management Tools   163

6.1   Standards   163

6.2   MIL-STD-499B Systems Engineering (Draft)   164
6.2.1   *General Standard Content*   164
6.2.2   *Systems Engineering Process*   166
6.2.3   *Content of the MIL-STD-499B SEMP*   166
6.2.4   *Additional Information and Requirements*   168
6.2.5   *Summary*   170

6.3   EIA/IS-632 Systems Engineering   171
6.3.1   *General Standard Content*   171
6.3.2   *Systems Engineering Process*   171
6.3.3   *Content of the EIA/IS-632 SEMP*   172
6.3.4   *Other Information and Requirements*   172
6.3.5   *Summary*   172

6.4   IEEE 1220 (Trial Use) and IEEE 1220-IEEE
        Standard for Application and Management of
        the Systems Engineering Process   173
6.4.1   *General Standard Content*   173
6.4.2   *IEEE 1220 Life-Cycle Model*   175
6.4.3   *Systems Engineering Process*   177
6.4.4   *Content of the IEEE 1220 Engineering Plan*   178
6.4.5   *Additional Material and Requirements*   179
6.4.6   *Summary*   179

6.5   ANSI/EIA-632-Processes for Engineering a System   179
6.5.1   *ANSI/EIA-632 Processes*   180
6.5.2   *ANSI/EIA-632 Requirements*   181

6.5.3   *ANSI/EIA-632 Concepts*   183

6.5.4   *ANSI/EIA-632 Annexes*   186

6.5.5   *Summary*   187

6.6   Other Useful Documents   187

6.6.1   *Technical Reviews and Audits*   188

6.6.2   *Systems Engineering Standards*   188

6.6.3   *Configuration Management*   188

6.6.4   *Specification Standards*   189

6.6.5   *Work Breakdown Structures*   189

6.7   Capability Maturity Models   189

6.8   SEI—Systems Engineering Capability Maturity Model   191

6.8.1   *SE-CMM Foundation*   192

6.8.2   *Process Areas*   192

6.8.3   *Capability Levels*   192

6.8.4   *Summary*   194

6.9   CMM Integration   194

Endnotes   196

7   **Systems Engineering Process Tools**   **199**

7.1   Analysis Tools—Requirements Engineering   199

7.1.1   *What Is a Requirement?*   200

7.1.2   *Requirements Engineering*   200

7.1.3   *Requirements Documentation*   209

7.1.4   *Automated Requirements-Management Tools*   210

7.1.5   *Difficulties in Developing Requirements*   212

7.2   Synthesis—Various Tools   214

7.2.1   *Schematic Block Diagrams*   214

7.2.2   *Physical Modeling*   215

7.2.3   *Mathematical Modeling and Simulation*   216

7.3   Evaluation—Trade-off Analysis   217

Endnotes   222

8   **Related Disciplines**   **225**

8.1   Introduction   225

8.2   Project Management   225

8.3   Quality Assurance   230

8.4   Logistics Support   231

8.5   Operations   233

8.6   Design Support Network   233

8.7   Software Engineering   234

8.8   Hardware Engineering   235

    Endnotes   235

List of Acronyms   **237**

About the Authors   **243**

Index   **245**

Preface

The need to manage complexity is now commonplace in almost all fields of undertaking. Complex systems such as cars, airplanes, airports, financial systems, and communications networks commonly involve millions of hours of work by thousands of people from a wide range of disciplines and backgrounds spread across a number of countries. Projects often take up to ten years to complete and involve a large number of disparate stakeholders, developers, operators, and customers. At the same time, the need to accommodate changes in the market place has created considerable pressure on traditional engineering processes. It is therefore little wonder that we have become used to hearing about the difficulties associated with complex projects—cost and schedule overruns, dramatic failures to achieve requirements, project cancellations, and so on.

Ensuring that each of the associated disciplines pays more attention to their profession does not solve these problems. Complex technical projects can only be managed effectively by addressing the whole life cycle. First, requirements must be formally defined to provide a comprehensive description of the functionality of the system to be procured—a functional architecture. These functional requirements are analyzed and elaborated upon to create a functional description of subsystem requirements, which are then allocated to physical configuration items to provide a physical architecture of the system. The aim of developing the physical configuration items is to reduce the complex system to a series of well-defined subsystems that can be designed and then built by manageable teams using extant processes and procedures. The subsequent development of these separate subsystems must be managed, however, so that they are verified, tested, and integrated into the final system to be delivered. To be successful, the entire process must be planned, documented, and managed.

This book provides a basic but complete coverage of the management of complex technical projects and, in particular, of the discipline known as *systems engineering* through which that management is conducted. We offer a framework encapsulating the entire systems engineering discipline, clearly showing where the multitude of systems engineering activities fits within the overall effort. The framework provides an ideal vehicle for understanding the complex discipline of systems engineering.

We take a top-down approach that introduces the philosophical aspects of the discipline and provides a framework within which the reader can assimilate the associated activities. Without such a reference, the practitioner is left to ponder the plethora of terms, standards, and practices that have been developed independently and often lack cohesion, particularly in nomenclature and emphasis. The field of systems engineering is often viewed as dry, detailed, complicated, acronym-intensive, and uninteresting. Yet the discipline holds the solution to delivering complex technical projects on time and within budget, avoiding many of the failures of the past. The intention of this book is both to cover all aspect of the discipline and to provide a framework for the consideration of the many issues associated with engineering complex systems.

Our secondary purpose is to describe a complex field in a simple, easily digested manner that is accessible to a wide spectrum of readers, from students to professionals, from novices to experienced practitioners. It is directed at a wide audience and aims to be a valuable reference for all professionals associated with the management of complex technical projects: project managers, systems engineers, quality assurance representatives, integrated logistic support practitioners, maintainers, and so on.

In line with the top-down approach of systems engineering, we focus in this book on the early stages of the system life cycle because the activities in these stages have the greatest impact on the successful acquisition and fielding of a system. In the interests of balance, however, we use the systems engineering framework to provide an overview of all other aspects related to systems engineering.

Chapter 1 is dedicated to the introduction of systems engineering and the development of the framework within which the remainder of the book is written. Chapters 2, 3, and 4 examine in more detail the issues associated with the most important activities undertaken during early design—that is, during the first three phases of the system life cycle (conceptual design, preliminary design, and detailed design and development). Emphasis is given to the early activities associated with conceptual and preliminary design, as these have the greatest impact on the system life cycle. Chapter 5 deals with the broad topic of systems engineering management and details some of the activities normally associated with engineering management. Some of the more common and popular systems engineering tools are introduced in Chapters 6 and 7, and their application to engineering management and process is explained. The final chapter explains the interrelationship between the systems engineering effort and other closely related disciplines, such as project management, quality management, and integrated logistics support management.

Systems engineering is a broad discipline, and its application to different projects always requires individual and independent thought. There is never a single solution that will work with all projects, and there is rarely a solution that is either completely right or wrong. This book aims to introduce the main systems engineering issues to the reader to facilitate some of that individual and independent thought.

1

Introduction to Systems Engineering

Systems engineering [1] methodologies and practices began to emerge from experience gained in the U.S. Department of Defense (DoD) acquisition programs of the 1950s. These programs often involved complex and challenging user requirements that tended to be incomplete and poorly defined. Additionally, most programs entailed high technical risk because they involved large numbers of different technical disciplines and the use of emerging technology. Following a number of program failures, the discipline of systems engineering emerged to help avoid, or at least mitigate, some of the technical risks associated with the complex equipment acquisition programs. Systems engineering provides a framework, within which complex systems can be adequately defined, analyzed, specified, manufactured, operated, and supported. Systems engineering processes and methodologies have continued to develop since the 1950s and are widely applied to many of today's challenging acquisition projects.

The focus of systems engineering is on the system as a whole and the maintenance of a strong interdisciplinary approach. Project management, quality assurance, integrated logistics support, and traditional design disciplines such as hardware and software engineering are but a few of the many disciplines that are part of a coordinated systems engineering effort.

Throughout the following chapters, we use a number of examples wherever possible to illustrate and reinforce the systems engineering theory being introduced. To avoid duplication and assist further with an understanding of the whole systems engineering process, throughout the chapters we also use a single worked example, based on the acquisition of an aircraft

1

system. We do not intend to replicate the design process for a modern aircraft and supporting elements. Rather, an aircraft system has been chosen as a convenient example that can be readily recognized by readers from a wide variety of disciplines and specialties. The illustrations are designed so that readers are not forced to become domain experts in a particular field just to understand the illustration. With the aircraft example, the majority of readers can immediately understand the system context, the need, the functional and performance requirements, the interface issues, technical performance measures, the functional-to-physical translation, broad trade-off analyses, and the physical configuration items involved in the final design. It should be noted, however, that we do not at all suggest that the aggregation of examples throughout the text represents an adequate design for an aircraft system; the available space prohibits the inclusion of sufficient detail, which would also obscure the general lessons that are to be illustrated by the example.

Example 1.1: Introduction to the Aircraft Example

A large aircraft operator (ACME Air) has identified a need for a medium-sized aircraft to replace the aging platform that it currently operates over domestic routes and some short international routes. The company wants to use a systems engineering approach to ensure that the aircraft system produced is ideally suited to the role and to ensure the overall commercial viability of the project.

This brief introduction provides the seed of an idea that grows throughout the remaining chapters as we consider the relevant systems engineering activities required to see the system through definition and design, construction and production, operation, maintenance, support, phaseout, and disposal.

1.1   What Is a System?

A *system* is a complex set of many often-diverse parts subject to a common plan or serving a common purpose. In the broadest sense, a system is something that provides a solution to a complex problem. With this in mind, a system combines a number of resources together in an organized manner so as to perform a collection of specified functions to specified levels of performance.

A system can be defined in two broad ways—in functional terms and in physical terms. A functional description of a system articulates what the system will do, how well it will do it, under what conditions it will perform, and what other systems will be involved with its operation. A physical description relates to the system components and explains what the components are, how they look, and how the components are to be manufactured and integrated. The two descriptions live independently as valid descriptions of a system, but an understanding of the relationship between functional and physical description leads to a deeper appreciation of the system.

A system must meet some defined need, goals, and objectives, which must be clearly stated by the user, and represent the starting point of the design process as well as the ultimate test of the system's fitness for purpose once it has been introduced into service.

A system is much more than an aggregation of hardware or software and must be described in terms of such resources as personnel, materials, facilities, data, hardware, and software. Analysis of these resources yields the majority of the system's functional and performance requirements. The system is fully defined by the combination of these resources operating in a live environment, which defines the context within which the system must operate and often dictates functional or performance requirements to be exhibited by the system.

Example 1.2: System Resources for Our Aircraft Example

Resources for our aircraft system example could include, but are not limited to the following:

- Personnel. *Air crew are required to operate the system and ground crew are required to maintain and support the fleet of aircraft.*

- Materials. *Materials are required to operate the system, including fuel, lubricants, and other consumables such as tires and spare parts.*

- Facilities. *The aircraft needs a network of maintenance facilities for routine maintenance and repairs throughout its life. Other facilities such as terminals are also necessary to operate the aircraft.*

- Data. *Data is required to maintain and operate the aircraft. Data could include maintenance information such as specifications and drawings, and operational information such as user manuals and instructions.*

- Hardware. *The most tangible part of the system is the hardware itself. The aircraft will be produced, distributed, and sold to operators who will then use the aircraft in a number of different ways, such as domestic and international operations for passengers or freight.*

- Software. *Software is rapidly becoming a critical item within many systems. The aircraft is likely to use hardware and software to control a range of functions from engine management, through navigation and environmental control systems, to the communications and flight-control systems.*

Throughout this book we use a number of terms to refer to the different parties involved in system development.

The customer organization is generally represented by an acquisition element (typically headed by a project manager) that procures the new system on behalf of the users (operators) within the organization who own the perceived need. In addition to the acquisition staff and users, there are many others within the customer organization who have a stake in the successful implementation of the project. These stakeholders include representatives from the management, financial, operations, supply, maintenance, and facilities areas, to name just a few.

The contractor is the entity responsible for designing and developing the system to meet the customer requirements. The relationship between the customer and the contractor varies with each project but, for each project, is defined by the terms and conditions of the contract. In most cases the contractor is not able to perform all of the work required and devolves packages of work to a number of subcontractors. The terms and conditions relating to this work are described in the subcontract.

Systems are hierarchical by nature. In this text, we limit our discussions to a three-layer hierarchy to describe the focus of various systems engineering processes. We describe a top-level entity known as the system that comprises a number of subsystems that, in turn, consist of a number of components. We could go further, but it serves little purpose in this text.

The application of the terms to specific situations and examples depends very much on the context of the situation and where within the overall picture the players involved in the situation sit. For example, if we talk about the highest-level considerations of the aircraft example, we would talk about the aircraft system consisting of, among others, the engine subsystem and the avionics subsystem. The engine subsystem may consist of such components as fuel tanks, pumps and lines, turbines, compressors and gear boxes, and hydraulic pumps. If, however, we were talking from the viewpoint of an engine manufacturer, we would talk of the engine as the system, comprising fuel, power plant and hydraulic subsystems, and so on.

1.2 System Life Cycle

The life cycle of a system commences with the statement of a need and ends with disposal of the system. In between there are a number of system phases and activities, each of which builds on the results of the preceding phase or activity. There is no universally accepted agreement on how many phases exist in a system life cycle or what those phases are called. For the discussion in this book, we use the system phases defined by Blanchard and Fabrycky [2] and MIL-STD-499B [3], and illustrated in Figure 1.1.

We have selected this model for a number of reasons. It emphasizes that a system begins with a perceived need and finishes upon disposal—the so-called cradle-to-grave approach. There is a clear delineation between the acquisition and in-service (utilization) phases of a system, allowing the application of systems engineering during utilization to be investigated and documented. In addition, it shows sufficient detail in the early stages of the acquisition, where the application of systems engineering methodologies and practices has the potential to make the most significant contribution.

As shown in Figure 1.1, the system life cycle can be divided into two very broad phases: the acquisition phase and the utilization phase. The life cycle of a system begins with a perceived need by a group of users, which provides the input for the acquisition phase that continues until the system is brought into service. The system stays in service until the utilization phase concludes with disposal of the system. Quite often the conclusion of one system life cycle marks the beginning of another system's life cycle.

The significance of focusing on the system life cycle is that decisions made early in the acquisition phase are informed of the proposed and intended activities in the utilization phase. For example, the design of an

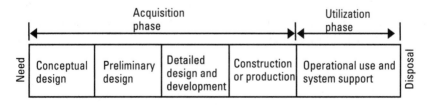

Figure 1.1 System life cycle.

aircraft airframe must take into account the maintenance and operation of that airframe during the utilization phase. It would be pointless to design the best airframe in the world if it did not have the necessary access points to allow maintenance personnel to service it or operators to operate it in the intended environment. Other examples of utilization phase requirements that have an impact on equipment design or selection during the acquisition phase include reliability and availability. Reliability normally refers to the ability of the equipment to operate without failure for a given period of time. Availability is a measure of the degree to which a system is in an operable condition when required at some random point in time. Again, a superior design is pointless unless the system can meet specified minimum levels of reliability and availability.

Economic factors provide arguably the most compelling evidence to support the focus on life cycle as opposed to product. In short, a life-cycle focus can save money in the long term. Experience has shown that a large proportion of total life-cycle cost (LCC) for a given system stems from decisions made early in the acquisition phase of the project. Some 60% of errors in system development originate in the requirements analysis process [4]. To that end, the maximum opportunity to reduce the total LCC of a system is presented in the acquisition phase. Poor requirements cannot be rectified by good design, so that it invariably follows that rigorous development of requirements is essential for the acquisition is to be successful.

The life cycle illustrated in Figure 1.1 shows the phases and activities in sequence and is not intended to represent any particular process model such as the waterfall, evolutionary development, incremental development, reusable components, technology application, reverse engineering, spiral, vee, or evolutionary acquisition models [5]. These models represent different approaches to implementing the activities of the system life cycle in Figure 1.1.

1.2.1 Acquisition Phase

Figure 1.1 shows that the acquisition phase comprises four main activities: conceptual design, preliminary design, detailed design and development, and construction and production.

1.2.1.1 Conceptual Design

The initial systems engineering effort is aimed at producing a clearly defined set of user requirements at the system level. This initial effort is referred to as *conceptual design* and represents the efforts to articulate the system design in functional terms. Although clearly defining the functional requirements of the system would seem a logical (and essential) first step, it is often poorly done and is most often the direct cause of problems later in the development process. Customers sometimes prefer to describe their needs in a loose and ambiguous manner to protect themselves from changes and developments in their needs. The conceptual design process aims to avoid this ambiguity and establishes what is called a *functional baseline* (which describes the whats and whys of the system). The functional baseline represents a system-level functional architecture that meets the customer needs. Conceptual design can therefore be considered to be another term for functional design.

1.2.1.2 Preliminary Design

With the initial functional baseline established, the preliminary design process can commence. The aim of preliminary design is to convert the functional baseline into a preliminary definition of the system configuration or architecture (the hows of the system). The preliminary definition of the system configuration represents the initial attempt at physical design. Preliminary design is therefore the stage where functional design is translated into physical design. This translation occurs through an iterative process of requirements analysis, synthesis (or design), and evaluation. The result of the preliminary design process is a subsystem-level design known as the *allocated baseline*. Note that the allocated baseline indicates that the functional requirements (defined in the functional baseline) have now been grouped together logically and allocated to subsystem-level components, which combine to form the overall system design. The allocated baseline therefore represents a subsystem-level physical architecture that meets the needs of the functional architecture in the functional baseline. Traceability between the functional and physical designs is a critical cornerstone of systems engineering that must be established and maintained during preliminary design and subsequent stages.

1.2.1.3 Detailed Design and Development

Detailed design and development is the next activity of the acquisition phase. The allocated baseline developed during preliminary design is used in the detailed design and development process to commence development of the individual subsystems and components in the system. Prototyping may occur and the system design is confirmed by test and evaluation. The result of the detailed design and development process is the initial establishment of the product baseline, as the system is now defined by the numerous products (subsystems and components) making up the total system. The definition of the system at this stage should be sufficiently detailed to commence the construction and production activities.

1.2.1.4 Construction and Production

The final activity within the acquisition phase is construction and production. The product baseline produced during detailed design and development will have been refined and finalized prior to entering this phase due to the results of the test and evaluation effort. System components will be produced in accordance with the detailed design specifications and the system is ultimately constructed in its final form. Formal test and evaluation activities will be conducted to ensure that the final system configuration meets its intended purpose. Configuration management activities called *configuration audits* will confirm that the system (as produced) agrees with the documentation comprising the product baseline prior to full-scale production occurring. At the successful completion of the configuration audits, the product baseline for the system is said to be approved or in place.

1.2.2 Utilization Phase

On delivery, the system moves into the utilization phase (the final process within the system life cycle prior to disposal). The major activities during this phase are operational use and system support. Systems engineering activities may continue during the utilization phase to support any modification activity that may be required. Modifications may be necessary to rectify performance shortfalls, to meet changing operational requirements or external environments, or to enhance current performance or reliability. Another common reason for modifications is to enable ongoing support for the system to be maintained.

Following operational use and system support, the system is eventually phased out and retired from service, completing the system's entire life cycle, which may take many years.

Example 1.3: The RAAF F-111 Aircraft

An excellent example of an extended life cycle is the F-111 aircraft currently being operated by the Royal Australian Air Force (RAAF). Following a midlife upgrade in the 1990s involving both avionics and engine modifications, the RAAF expects the F-111 to remain in service until 2020. Because the origins of the F-111 can be traced back to the 1950s, the life cycle of the RAAF F-111s may eventually exceed 70 years.

1.3 What Is Systems Engineering?

There is a wide range of systems engineering definitions, each of which tends to reflect the particular focus of its source. The following are some of the more accepted and authoritative definitions of systems engineering from recent standards and documents.

> Systems engineering is the management function which controls the total system development effort for the purpose of achieving an optimum balance of all system elements. It is a process which transforms an operational need into a description of system parameters and integrates those parameters to optimize the overall system effectiveness. [6]

> An interdisciplinary collaborative approach to derive, evolve, and verify a life-cycle balanced system solution which satisfies customer expectations and meets public acceptability. [7]

> An interdisciplinary approach encompassing the entire technical effort to evolve and verify an integrated and life cycle balanced set of system, people, product, and process solutions that satisfy customer needs. Systems engineering encompasses: the technical efforts related to the development, manufacturing, verification, deployment, operations, support, disposal of, and user training for, system products and processes; the definition and management of the system configuration; the translation of the system definition into work breakdown structures; and development of information for management decision making. [8]

> Systems engineering is the selective application of scientific and engineering efforts to: transform an operational need into a description of the system configuration which best satisfies the operational need according to the measures of effectiveness; integrate related technical parameters and ensure compatibility of all physical, functional, and technical program interfaces in a manner which optimizes the total system definition and design; and integrate the efforts of all engineering disciplines and specialties into the total engineering effort. [9]

> Systems engineering is an interdisciplinary, comprehensive approach to solving complex system problems and satisfying stakeholder requirements. [10]

Although each of these definitions has a slightly different focus, a number of common themes are evident and are described in the following sections.

1.3.1 Requirements Engineering

The complete and accurate definition of system requirements is a primary focus of the early systems engineering effort. The life cycle of a system begins with a simple statement of need, which is translated into a large number of statements of requirement that form the basis for the functional design and subsequently the physical architecture. These transitions must be managed by a rigorous process that guarantees that all relevant requirements are included (and all irrelevant requirements excluded). The establishment of correct requirements is fundamental to the success of the subsequent design activities.

Once requirements have been collected, the systems engineering process then focuses on the management of these requirements from the system level right down to the lowest constituent component. This requirements engineering (sometimes referred to as *requirements management* or *requirements flow down*) involves elicitation, analysis, definition, and validation of system requirements. Requirements engineering (described in more detail in Section 7.1) ensures that a rigorous approach is taken to the collection of a complete set of unambiguous requirements from the stakeholders.

Requirements traceability is also an essential element of effective management of complex projects. Through traceability, design decisions can be traced from any given system-level requirement down to a detailed design decision (forward traceability). Similarly, any individual design decision must be able to be justified by being associated with at least one higher-level requirement (backwards traceability). This traceability is important because the customer must be assured that all requirements can be traced forward and can be accounted for in the design at any stage. Further, any aspect of the design that cannot be traced back to a higher-level requirement is likely to represent unnecessary work for which the customer is most probably paying a premium. Traceability also supports the change process, especially the investigation of change impact.

Support for requirements traceability is a feature of the top-down approach that provides a mechanism by which it can be guaranteed that requirements can be satisfied at any stage. A bottom-up approach cannot provide the same guarantee.

1.3.2 Top-Down Approach

Traditional engineering design methods are based on bottom-up approach in which known components are assembled into subsystems from which the system is constructed. The system is then tested for the desired properties and the design is modified in an iterative manner until the system meets the desired criteria. This approach is valid and extremely useful for relatively straightforward problems that are well defined. Unfortunately, complex problems cannot be solved with the bottom-up approach.

Systems engineering begins by addressing the complex system as a whole, which facilitates the initial allocation of requirements as well as the subsequent analysis of the system and its interfaces. Once system-level requirements are understood, the system is then broken down into subsystems and the subsystems further broken down into components until a complete understanding is achieved of the system from top to bottom. This top-down approach is a very important element of managing the development of complex systems. By viewing the system as a whole initially and then progressively breaking the system into smaller elements, the interaction between the components can be understood more thoroughly, which assists in identifying and designing the necessary interfaces between components (internal interfaces) and between this and other systems (external interfaces). For example, Figure 1.2 illustrates the ANSI/EIA-632 approach to top-down development [11].

It must be recognized, however, that while design is conducted top down, the system is implemented using a bottom-up approach. That is, the aim of system engineering can be considered to be to provide a rigorous, reproducible process by which the complex system can be broken into a series of simple components that can then be designed using the traditional engineering bottom-up approach. Importantly, the other facet of systems engineering is to provide a process by which the components and subsystems can be integrated to achieve the desired system properties.

Note that, as discussed in Section 1.1, the terms *system, subsystem,* and *component* are relative. Each system comprises subsystems that consist of components. Each subsystem, however, can be considered to be a system in its own right, which has subsystems and components and so on. Consequently, while the building-block concept for top-down development is very useful, it often provides for confusion among novices due to the relative nature of the associated terms.

1.3.3 Focus on Life Cycle

Systems engineering is focused on the entire system life cycle and takes this life cycle into consideration during decision-making processes. As described in Section 1.2, a system's life cycle begins during system definition and design, and passes through construction and production, operation,

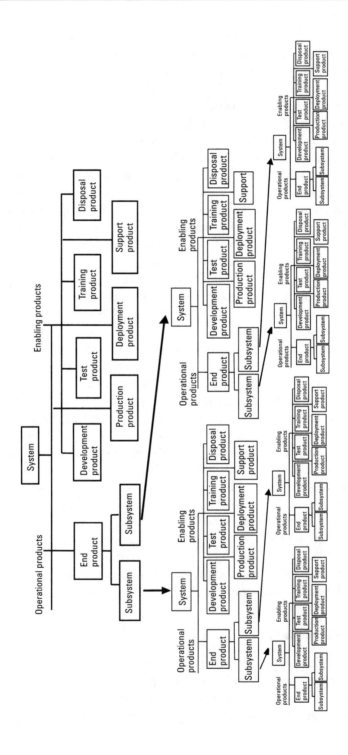

Figure 1.2 ANSI/EIA-632 building-block concept for top-down development [11].

maintenance, support, and phaseout. The life cycle concludes only with the disposal of the system.

In the past it was common to consider design options only in the light of the issues associated with the acquisition phase and to pay little attention to through-life support issues. Project teams typically focus on the acquisition phase of the project and on the development of a system that meets the functional user requirements while minimizing cost and schedule. This has often led to larger-than-expected costs in the utilization phase to be met from budgets that are insufficient to keep systems in service. A life-cycle focus requires a system focus not a product focus. A system focus takes into account all constituent elements of the system, including operation, maintenance and support, and retirement or disposal.

Example 1.4: Life-Cycle Focus

As a simple analogy to demonstrate the concept (and problems) of a product focus as opposed to a life-cycle focus, consider a motor vehicle manufacturer that has introduced a new model. The new family sedan is a roomy, good-looking, high-performance vehicle that costs half the price of similar models. Strong interest seems assured with buyers attracted by the vehicle's features as well as the cost-effectiveness of the purchase.

On further investigation, however, it transpires that the low purchase price has been achieved by a design that incorporates poor-quality components that require iterative, regular maintenance resulting in running costs that are five times higher than competing models.

Clearly, a decision to purchase the new vehicle must focus on more than just the purchase price of the vehicle and must take into account the system life-cycle aspects of operation, maintenance, and support.

1.3.4 System Optimization and Balance

As we discuss in Chapter 3, it does not necessarily follow that the combination of optimized subsystems leads to an optimized system. Additionally, the system architecture must represent a balance between the large number of requirements, that cover a wide range of factors such as environmental, moral, ethical, social, cultural, psychological, and economic human factors, in addition to the technical considerations. A further advantage, therefore, of the top-down approach in systems engineering is that system optimization and balance can be achieved as a byproduct of the design process, something that cannot be guaranteed in a bottom-up design method.

A balance must also be struck across the life cycle. Metrics such as cost-effectiveness must be measured across all phases, not just acquisition.

1.3.5 Integration of Disciplines and Specialties

Systems engineering aims to manage and integrate the efforts of a multitude of technical disciplines and specialties to ensure that all user requirements are adequately addressed. Rarely is it possible for a complex system to be designed by a single discipline. Consider our aircraft example. While aeronautical engineers may be considered to have a major role, the design, development, and production of a modern aircraft system requires a wide variety of other engineering disciplines including electrical, electronics, communications, radar, metallurgical, and corrosion engineers. Of course, in system terms, other engineering disciplines are required for testing and for logistics and maintenance support as well as the design and building of facilities such as runways, hangars, refueling facilities, and embarkation and disembarkation facilities. Other nonengineering disciplines are involved in marketing, finance, accounting, legal, environmental, and so on. In short, there could be hundreds, even thousands, of engineers and members of other disciplines involved in the delivery of an aircraft system.

The aim of the systems engineering function is to break up the task into components that can be developed by these disparate disciplines and specialties and then provide the management to integrate their efforts to produce a system that meets the users requirements. In modern system developments, this function is all the more important because of the complexity of large projects and their contracting mechanisms, and the geographic dispersion of contractor and subcontractor personnel across the country and around the world.

1.3.6 Management

While systems engineering clearly has a technical role and provides essential methodologies for systems development, it is not limited simply to technical issues and is not simply another engineering process to be adopted. As we discuss throughout this text, systems engineering has both a management and a technical role. Project management is responsible for ensuring that the system is delivered on time, within budget, and meets the customers expectations. The trade-offs and compromises implicit in those functions are informed by the products of systems engineering. Systems engineering and project management are therefore inextricably linked. These issues are discussed in more detail in Chapter 8.

1.4   Systems Engineering Relevance

Systems engineering principles and processes are applicable (albeit to varying degrees) to a wide range of projects. For example, EIA-632 [12] states that the standard itself is intended to be applicable to

the engineering or the reengineering of

a) commercial or non-commercial systems, or part thereof;

b) any system, small or large, simple or complex, software-intensive or not, precedented or unprecedented;

c) systems containing products made up of hardware, software, firmware, personnel, facilities, data, materials, services, techniques, or processes (or combinations thereof);

d) a new system or a legacy system, or portions thereof.

Because it is difficult to imagine a project that does not fit into this description, it is critical to understand the merits of systems engineering and apply them in a tailored manner, cognizant of the relative size, complexity, and risks associated with each undertaking. At one end of the spectrum are large complex projects making use of leading-edge developmental technology. These projects typically involve large sums of money, long time scales, and significant risks. At the other end of the spectrum are small projects making use of extant techniques and existing technology. These projects typically involve short periods, low costs, and a minimum of risk. Clearly different levels of systems engineering are applied to each of these types of projects.

The most obvious application of systems engineering principles and methodologies is in projects that are large and complicated. However, smaller and less complex projects can also benefit from the application of systems engineering principles.

Example 1.5: Systems Engineering Relevance to Large Projects

A classic example cited by the Software Engineering Institute in their Systems Engineering Capability Maturity Model (SE-CMM), Version 1.0 [13] demonstrates the need for good systems engineering. The Tacoma

Narrows Bridge was a very long suspension bridge with a flexible roadway that collapsed in 1940 due to strong winds that had set up an aerodynamic oscillation.

It is believed that the problems associated with the road could have been avoided or managed more effectively had systems engineering principles and processes been applied rigorously during the project. While adhering to good systems engineering practices would not necessarily ensure that designers are more likely to know the parameters of a particular problem, rigorous developmental approaches are more likely to draw out the relevant risk areas of a system (such as environment-related risks in this case).

The failures of system developments are often attributed to misunderstandings, ambiguities, misinterpretations, errors, and omissions in what the contractor is attempting to deliver. The result is a system that fails to solve the customer's problem, leading to a breakdown in the relationship between contractor and customer. Systems engineering specifically targets these problems and therefore is relevant to both parties in any system development.

Within a customer's project office, systems engineering is particularly relevant to the technical personnel who are responsible for the application of systems engineering principles as part of the overall project management effort. In most customer organizations, systems engineering personnel report to the project manager. The systems engineer is in an excellent position to apply the tools of systems engineering to assist the project manager in each of the nine project management knowledge areas (discussed in more detail in Chapter 8). Systems engineering is therefore an essential element of the project manager's ability to acquire a quality system within budget, time, and scope constraints.

Contractors should also be focused on developing processes and best practices to support the delivery of superior products and services to their customers. Systems engineering offers tools and philosophies that support the consistent development of quality products and services.

1.5 Systems Engineering Benefits

There are a number of potential benefits to be gained from the successful implementation of systems engineering processes and methodologies.

The first and most visible benefit is the scope for saving money during all phases of the system life cycle—LCC savings. While some may argue that the additional requirements imposed by systems engineering can increase costs, these increases are generally felt in the very early design phases. If applied appropriately, systems engineering can ensure that the savings achieved far outweigh the cost of implementing appropriate procedures and methodologies. Experience indicates that an early emphasis on systems engineering can result in significant cost savings later in the construction and production, operational use and system support, and disposal phases of the life cycle [14].

Systems engineering should also assist in reducing the overall schedule associated with bringing the system into service. Systems engineering ensures that the user requirements are accurately reflected in the design of the system helping to minimize costly and time-consuming changes to requirements later in the life cycle. If changes are required, they can be incorporated early in the design and in a controlled manner. The rigorous consideration and evaluation of feasible design alternatives during the design phases of the project promote greater design maturity earlier.

System failures, cost overruns, and schedule problems are often the direct result of poor requirements-management practices. The systems engineering discipline aims to put in place a rigorous process of requirements management to produce well-defined requirements, adequate levels of traceability between the different levels of technical design documentation back to the original user requirements, and requirements which are both verifiable and consistent. This requirements-management process must achieve these results without presupposing a particular technical solution or placing unnecessary technical constraints on the solution.

Figure 1.3 illustrates the impact of systems engineering on the system life cycle. For more complex system developments, the profile may be more complex than Figure 1.3, especially when the development spans many years and early production systems are used to refine the functionality and performance of later versions. Note that systems engineering has its greatest impact through the rigorous application of processes and methodologies during the early stages of the project, where the ease of change and cost of modification is the lowest. In fact, the curve in Figure 1.3 could be relabeled as the ease with which changes can be made throughout the system life cycle. Because systems engineering has the largest impact during the early stages of the process, in this book we concentrate primarily on the conceptual, preliminary, and early detailed design processes.

Systems engineering leads to a reduction in the technical risks associated with the product development. Risks are identified early and monitored throughout the process using a system of technical performance measures, design reviews, and audits. Design decisions can be traced back to the original

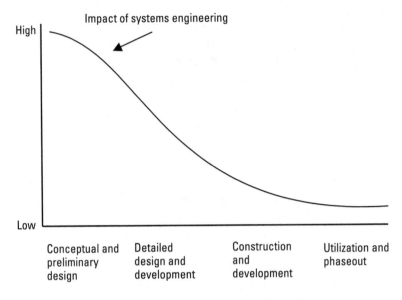

Figure 1.3 Impact of systems engineering on the system life cycle.

user requirements and conflicting user requirements can be identified and clarified early, significantly reducing the risk of failure later in the project.

Finally, the disciplined approach to requirements engineering leads to a product that meets the original intended purpose more completely. This improved functional performance makes for a quality system, where quality is measured by the ability of the system to meet the documented requirements.

1.6   Analysis, Synthesis, and Evaluation

All extant systems engineering standards and practices extol processes that are built around an iterative application of analysis, synthesis, and evaluation. The iterative nature of the application is critical to the systems engineering processes. Initially the process is applied at the systems level. It is then reapplied at the next level of detail and so on until the entire development process is complete. During the earlier stages the customer is heavily involved; in the latter stages, the contractor is mainly responsible for the continuing effort, which is monitored by the customer.

Prior to detailing the individual activities within the systems engineering processes, it is worth considering the basic foundations of the analysis-synthesis-evaluation loop illustrated in Figure 1.4. This concept is neither new nor complex; it is simply a good, sound approach to problem solving. While applicable in any domain, the loop is particularly fundamental to systems engineering.

1.6.1 Analysis

Analysis commences with a statement of perceived need or a set of customer requirements. During conceptual design, analysis investigates these needs and requirements and identifies the essential functions that the system must perform in order to meet the needs. Requirements analysis at the system level aims to answer the what, how well, and why questions relative to the system design. Analysis activities continue throughout the subsequent stages of the life cycle to help in defining lower level requirements, often called *derived requirements,* associated with physical aspects of system design.

A basic tool that can be used during analysis is the functional flow block diagram (FFBD), which shows the logical sequences and relationships of the functions of the system at the system level. The system-level functions detailed in the FFBD are further developed at the subsystem level and below using more detailed FFBDs. Examples of FFBDs are presented in Chapters 2 and 3. As the design progresses through the different design phases and becomes more detailed, the system-level functions will be individually investigated and have an FFBD produced and so on until a very detailed set of functions and their interrelationships have been identified.

Once all the functions associated with the system have been identified, the requirements associated with each function can be defined. These requirements could include performance parameters such as speed, altitude, and accuracy; interoperability requirements detailing other systems with which the system under development must operate; and interface requirements to describe the necessary outputs expected from the system and the

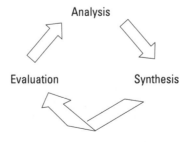

Analysis

Evaluation   Synthesis

Figure 1.4 The analysis-synthesis-evaluation iteration.

inputs to the system. Depending on the particular design phase, these requirements and functions may also be grouped in accordance with some sort of logical criteria and then allocated to a particular physical component of the system. That is, the component becomes responsible for the satisfaction of those requirements by performing the functions assigned to it. The allocation of requirements forms a description of the system elements and architecture and therefore assists in the process of synthesis or design (answering the how questions).

1.6.2   Synthesis

The analysis activity resolves what is required, as well as how well and why. Synthesis, or design, now determines how. Synthesis is possibly the most widely recognized role of a professional engineer. Synthesis is the process whereby creativity and technology are combined to produce a design that best meets the stated system requirements. The term synthesis is more appropriate than design in the systems engineering context, as it hints at the evolutionary nature of design and development.

In the early stages of the systems engineering processes, synthesis is limited to defining completely the functional design of the system and then considering all possible technical approaches using the results from the requirements analysis effort. From this consideration, the best approach is selected and the process moves to the next level of detail. Later in the systems engineering processes, the selected design concept is synthesized further until, ultimately, the complete system design is finalized.

Although synthesis is the creative part of the systems engineering effort, a number of tools are available to the engineer to ensure that all alternatives are considered and the most suitable alternative is ultimately selected. Some potential aids to assist the design engineer during the synthesis process are presented in Chapter 7.

1.6.3   Evaluation

System cost and risk are directly associated with requirements and design. Evaluation is the process of investigating the trade-offs between requirements and design, considering the design alternatives, and making the necessary decisions. The process of evaluation continues throughout all stages of the systems engineering effort, ultimately determining the system's satisfaction of original requirements. Trade-off analysis is one of the tools available to the system designer in performing evaluation of competing requirements or designs—a detailed treatment of trade-off analysis is provided in Chapter 7.

The outcome of the evaluation is a selection or confirmation of the desired approach to design. Discrepancies are also identified if applicable and may result in further analysis and synthesis as the analysis-synthesis-evaluation loop is closed.

1.7 A Systems Engineering Framework

Discussions on systems engineering become complicated by the broad mandate of the system, the complexity and interrelationship of the many systems engineering constituents, and the relationships with other disciplines throughout the entire system life cycle.

The ability to understand a complex subject such as systems engineering is greatly enhanced by a solid framework within which concepts can be considered. An excellent example is the Project Management Body of Knowledge (PMBOK) [15] that provides a clear framework within which to consider the many facets of project management. Without an equivalent framework, the broad scope of systems engineering soon becomes confusing, given the complexity of its components and their many interrelationships. There are a number of excellent systems engineering standards available today that contribute to the elements of a suitable framework, but each standard contains complexity, terminology, and detail requiring substantial interpretation. The entry level of many students, young engineers, and project managers therefore does not allow the use of such standards as effective frameworks within which to examine systems engineering.

A systems engineering framework [16] (illustrated in Figure 1.5) has been synthesized by the authors through a thorough survey of existing systems engineering publications and standards and via experience in teaching systems engineering at a range of levels. The main aim of the framework is to provide a framework within which the systems engineering discipline can be understood and implemented.

The framework illustrates the relationship of the three main elements of systems engineering processes, management, and tools and places them in context with related disciplines. In common use, the terms systems engineering management and systems engineering processes are sometimes used interchangeably. Here we make a distinction between the two. We present the engineering processes as being the hows of systems engineering, the application of which forms the foundation of the systems engineering effort.

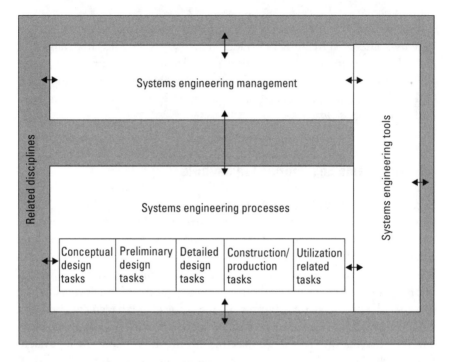

Figure 1.5 A framework for the consideration of systems engineering.

Over the top of these processes sits the systems engineering management function, which is responsible for directing the systems engineering effort, monitoring and reporting that effort to the appropriate areas, and reviewing and auditing the effort at critical stages in the entire process. These two elements are supported by a range of tools, and all elements interface with such related disciplines as traditional engineering, project management, integrated logistics support, and quality assurance.

The systems engineering framework provides an excellent structure within which to examine and explain the complex discipline of systems engineering. Experience in undergraduate and graduate courses as well as commercial short courses [17] has shown that the framework provides an excellent means of communicating the complexities and interrelationships of the systems engineering discipline, particularly to those who do not have a significant amount of project experience. Students can successfully grasp the fundamental concepts of systems engineering within a relatively short time frame.

1.7.1 Systems Engineering Processes

The basic systems engineering process is the analysis-synthesis-evaluation loop described in Figure 1.4. This loop is applied iteratively throughout the system life cycle. Other systems engineering processes and tasks are divided into the life-cycle stages within which they typically occur. In this book we do not attempt to detail exhaustively all systems engineering processes. Instead, we concentrate on the intent and main aim of each phase and examine some of the likely techniques that may be used to arrive at that aim. We place particular emphasis on the acquisition phase of the life cycle, as it is the phase during which systems engineering has the ability to have the most impact on a system.

In Chapter 2, for example, the tasks completed during conceptual design are shown to focus on achieving a clear and complete definition of the system-level requirements. We investigate the processes of articulation of the system needs, goals, and objectives; system feasibility analysis; system requirements analysis; system synthesis; and so on. Chapter 3 investigates the processes associated with preliminary design, and Chapter 4 describes those related to detailed design and development and construction and production.

1.7.2 Systems Engineering Management

Systems engineering management sits over the top of systems engineering processes and is responsible for directing the systems engineering effort, monitoring and reporting that effort to the appropriate areas, and reviewing and auditing the effort at critical stages in the entire process. In Chapter 5, we address the major systems engineering management elements of technical reviews and audits, system test and evaluation, technical risk management, configuration management, the use of specifications and standards, integration management, and systems engineering management planning.

The preeminent position of systems engineering management in the framework illustrates that it is the key to the entire systems engineering effort.

1.7.3 Systems Engineering Tools

Many tools exist to assist systems engineering processes and management. These tools range from techniques and methods to systems engineering standards. Here we describe the most popular tools and standards under the headings of management tools (Chapter 6) and process tools (Chapter 7). It is not the intended purpose of this book, however, to repeat information contained in standards and documents elsewhere.

Throughout the book we present generic process tools such as requirements breakdown structures, FFBD, work breakdown structures, trade-off analysis, prototyping, and simulation as examples of tools that may be applied to the systems engineering process effort. We also describe the systems engineering management tools of standards and capability maturity models. In Chapter 6, the current standards are reviewed and summarized including MIL-STD-499B [18], EIA/IS 632 [19], IEEE 1220 [20], and EIA 632 [21]. The SE-CMM [22] is used as an example capability maturity model (CMM) to describe how CMMs can be used to assist the overall systems engineering management effort.

1.7.4 Related Disciplines

There are many disciplines (both technical and nontechnical) related to systems engineering. Examples include project management, logistics management, quality assurance, user-requirements management, software engineering, hardware engineering, and interface engineering (or integration engineering). Related disciplines can be considered to be the glue that holds together the other components of the framework.

The relationship between the related disciplines and the other facets of systems engineering depends very much upon the discipline in question. Some (such as project management) oversee the whole systems engineering discipline, while others (such as hardware and software engineering) sit between systems engineering management and the processes, while others (such as quality assurance) sit alongside the systems engineering effort. Chapter 8 discusses these disciplines and their relationship with systems engineering.

Endnotes

[1]   Further reading on systems engineering can be obtained from the following major sources:

Aslaksen, E., and R. Belcher, *Systems Engineering,* Upper Saddle River, NJ: Prentice Hall, 1992.

ANSI/EIA-632-1998, *EIA Standard-Processes for Engineering a System,* Arlington, VA: Electronic Industries Association, 1999.

Beam, W., *Systems Engineering: Architecture and Design*, New York: McGraw-Hill, 1990.

Belcher, R., and E. Alaksen, *Systems Engineering*, Sydney: Prentice Hall, 1992.

Blanchard, B., *Systems Engineering Management*, New York: John Wiley & Sons, 1998.

Blanchard, B., and W. Fabrycky, *Systems Engineering and Analysis*, Upper Saddle River, NJ: Prentice Hall, 1998.

Boardman, J., *Systems Engineering: An Introduction*, Upper Saddle River, NJ: Prentice Hall, 1990.

Chestnut, H., *Systems Engineering Tools*, New York: John Wiley & Sons, 1965.

Chestnut, H., *Systems Engineering Methods*, New York: John Wiley & Sons, 1967.

Defense Systems Management College, *Systems Engineering Management Guide*, Washington, D.C.: U.S. Government Printing Office, 1990.

EIA/IS-632, *EIA Interim Standard-Systems Engineering*, Washington, D.C.: Electronic Industries Association, 1994.

Eisner, H., *Essentials of Project and Systems Engineering Management*, New York: John Wiley & Sons, 1997.

Forrester, J., *Principles of Systems*, Cambridge, MA: The MIT Press, 1968.

Gause, D., and G. Weinberg, *Exploring Requirements: Quality Before Design*, New York: Dorset House, 1989.

Grady, J., *System Engineering Planning and Enterprise Identity*, Boca Raton, FL: CRC Press, 1995.

Grady, J., *Systems Integration*, Boca Raton, FL: CRC Press, 1994.

Grady, J., *Systems Requirements Analysis*, New York: McGraw-Hill, 1994.

Hall, A., *A Methodology for Systems Engineering*, Princeton, NJ: D. Van Nostrand, 1962.

Hitchins, D., *Putting Systems to Work*, Chichester, England: John Wiley & Sons, 1992.

Hubka, V., and W. Eder, *Theory of Technical Systems*, Berlin: Springer-Verlag, 1988.

Hunger, J., *Engineering the System Solution: A Practical Guide to Developing Systems*, Upper Saddle River, NJ: Prentice Hall, 1995.

IEEE-STD-1220-1994, *IEEE Trial-Use Standard for Application and Management of the Systems Engineering Process*, New York: IEEE Computer Society, 1995.

IEEE-STD-1220-1998, *IEEE Standard for Application and Management of the Systems Engineering Process*, New York: IEEE Computer Society, 1998.

INCOSE, *J. Int. Council Syst. Engineering*, Seattle, WA: International Council on Systems Engineering.

INCOSE, *Proceeding of Annual Conference*, Seattle, WA: International Council on Systems Engineering.

Kotonya, G., and I. Sommerville, *Requirements Engineering: Processes and Techniques*, West Sussex, England: John Wiley & Sons, 2000.

Lacy, J., *Systems Engineering Management: Achieving Total Quality*, New York: McGraw-Hill, 1992.

Macaulay, L., *Requirements Engineering*, London, England: Springer-Verlag, 1996.

Machol, R., *System Engineering Handbook*, New York: McGraw-Hill, 1965.

Maier, M., and E. Rechtin, *The Art of Systems Architecting*, Boca Raton, FL: CRC Press, 2000.

Martin, J., *Systems Engineering Guidebook: A Process for Developing Systems and Products*, Boca Raton, FL: CRC Press, 1997.

MIL-STD-499B, *Military Standard—Systems Engineering—Draft*, Washington, D.C.: U.S. Department of Defense, 1994.

Rechtin, E., *Systems Architecting: Creating and Building Complex Systems*, Upper Saddle River, NJ: Prentice Hall, 1991.

Reilly, N., *Successful Systems Engineering for Engineers and Managers*, New York: Van Nostrand Reinhold, 1993.

Robertson, S., and J. Robertson, *Mastering the Requirements Process*, Harlow, England: Addison-Wesley, 1999.

Sage, A., *Decision Support Systems Engineering*, New York: John Wiley & Sons, 1991.

Sage, A., *Systems Engineering*, New York: John Wiley & Sons, 1992.

Sage, A., *Systems Management for Information Technology and Software Engineering*, New York: John Wiley & Sons, 1995.

Sage, A., and J. Armstrong, *Introduction to Systems Engineering*, New York: John Wiley & Sons, 2000.

SECMM-95-01, *Systems Engineering Capability Maturity Model*, Version 1.1, Carnegie Mellon University, Pittsburgh, PA: Software Engineering Institute, 1995.

Shishko, R., *NASA Systems Engineering Handbook*, Washington, D.C.: NASA, 1995.

Stevens, R., et al., *Systems Engineering: Coping with Complexity*, Hertfordshire, England: Prentice Hall, 1998.

Thome, B. (ed.), *Systems Engineering: Principles and Practice of Computer-Based Systems Engineering*, New York: John Wiley & Sons, 1993.

Truxal, J., *Introductory Systems Engineering Management*, New York: McGraw-Hill, 1972.

Young, R., *Effective Requirements Practices*, Boston, MA: Addison-Wesley, 2001.

Westerman, H., *Systems Engineering Principles and Practice*, Norwood, MA: Artech House, 2001.

Wymore, A., *Model-Based Systems Engineering*, Boca Raton, FL: CRC Press, 1993.

[2] Blanchard, B. and W. Fabrycky, *Systems Engineering and Analysis*, Upper Saddle River, NJ: Prentice Hall, 1998.

[3] MIL-STD-499B, *Military Standard-Systems Engineering-Draft*, Washington, D.C.: U.S. Department of Defense, 1994.

[4] Weinberg, J., *Quality Software Management. Volume 4: Anticipating Change*, New York: Dorset House, 1997.

[5] Further detail on these models can be found in the following, inter alia:

Blanchard, B., and W. Fabrycky, *Systems Engineering and Analysis*, Upper Saddle River, NJ: Prentice Hall, 1998, pp. 30–31.

Martin, J., *Systems Engineering Guidebook: A Process for Developing Systems and Products*, Boca Raton, FL: CRC Press, 1997.

[6] Defense Systems Management College, *Systems Engineering Management Guide*, Washington, D.C.: U.S. Government Printing Office, 1990.

[7] IEEE-STD-1220-1994, *IEEE Trial-Use Standard for Application and Management of the Systems Engineering Process*, New York: IEEE Computer Society, 1995.

[8] EIA/IS 632, *Systems Engineering*, Washington, D.C.: Electronic Industries Association, 1994.

[9] SECMM-95-01, *Systems Engineering Capability Maturity Model*, Version 1.1, Carnegie Mellon University, Pittsburgh, PA: Software Engineering Institute, 1995.

[10] Lake, J., *Unraveling the Systems Engineering Lexicon*, Proceedings of the INCOSE Symposium, 1996.

[11] ANSI/EIA-632-1998, *Processes for Engineering a System*, Washington, D.C.: Electronic Industries Association, 1999.

[12] ANSI/EIA-632-1998, *Processes for Engineering a System*, Washington, D.C.: Electronic Industries Association, 1999.

[13] CMU/SEI-94-HB-04, *A Systems Capability Maturity Model*, Version 1.0, Pittsburgh, PA: Software Engineering Institute, Carnegie Mellon University, 1994.

[14] Blanchard, B., and W. Fabrycky, *Systems Engineering and Analysis*, Upper Saddle River, NJ: Prentice Hall, 1998.

[15] *PMBOK, A Guide to the Project Management Body of Knowledge*, Upper Darby, PA: Project Management Institute, 1996.

[16] Faulconbridge, R., *Systems Engineering Body of Knowledge*, Canberra, Australia: Magpie Applied Technology, 2000.

[17] Faulconbridge, R., "A Systems Engineering Framework," *J. Battlefield Tech.*, Vol. 3, No. 2, July 2000.

[18] MIL-STD-499B, *Military Standard—Systems Engineering–Draft*, Washington, D.C.: U.S. Department of Defense, 1994.

[19] EIA/IS 632, *Systems Engineering*, Washington, D.C.: Electronic Industries Association, 1994.

[20] IEEE-STD-1220-1998, *IEEE Standard for Application and Management of the Systems Engineering Process*, New York: IEEE Computer Society, 1998.

[21] ANSI/EIA-632-1998, *Processes for Engineering a System*, Washington, D.C.: Electronic Industries Association, 1999.

[22] SECMM-95-01, *Systems Engineering Capability Maturity Model*, Version 1.1, Carnegie Mellon University, Pittsburgh, PA: Software Engineering Institute, 1995.

2

Conceptual Design

2.1 Introduction

The systems engineering processes begin very simply with the identification of a need for a new or improved system. The first activity in the acquisition phase is conceptual design, which is perhaps the most critical of all of the activities because it is responsible for the expansion of the system definition from a short statement of need into a functional architecture that may be hundreds of pages long.

The principal aims of conceptual design, therefore, are to articulate the need, to analyze and document the system-level requirements flowing from the need, and to complete a functional design of the system. The major product of the conceptual design, called the *functional baseline*, provides a system-level functional architecture that is the basis for subsequent lower level design.

Conceptual design is normally the domain of the customer, who is responsible for and heavily involved in this activity. While other organizations may be involved, the functional design of the system is normally considered to be the responsibility of the customer who must determine what the system needs to do and how well it needs to do it.

Figure 2.1 illustrates that five major tasks are conducted during conceptual design, including articulation of the stakeholder requirements, system feasibility analysis, system requirements analysis, system synthesis, and the system design review. These tasks are described in more detail in the following sections.

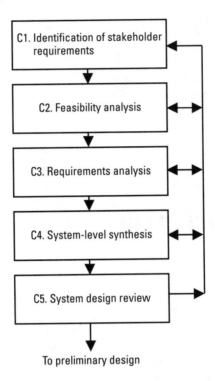

Figure 2.1 The five major tasks performed during conceptual design.

2.2 Identify Stakeholder Requirements

Before any work can commence on developing the system, the basic stakeholder requirements must be clearly and completely articulated. It would appear obvious that an idea of what needs to be achieved must be clearly understood and articulated before any further work is undertaken. Yet, a surprising number of system developments commence without a clear and complete understanding of the fundamental requirements. Not surprisingly most of these developments founder or, at the very least, are introduced into service with poor levels of user acceptance.

2.2.1 Stakeholder-Requirements Document

The first critical document in the systems engineering process captures the stakeholder requirements. This document is an essential first step towards a successful system development and is given a number of titles, including the operational concept document or description (OCD), concept of operations document (COD), system design document (SDD), user requirements document (URD), or simply user requirements. At the risk of further confusing the issue, we refer to the customer's document as the stakeholder requirements document (SRD) to avoid the use of a term that may belong to a particular process and to emphasize that requirements at this level must extend well beyond the operational requirements and capture other aspects of the system such as the logistics support concept. In other words, all stakeholders must be involved during the preparation of the SRD.

A reader of the SRD should be able to understand completely the likely applications or missions for which the system is intended, the major operational characteristics to be exhibited by the system, the operational constraints that limit the design and development of the system, the external systems and interfaces with which the system under development must operate, the operational and support environment within which the system must exist, and the support concept to be employed to support the system and enable it to continue performing in accordance with customer expectations [1].

The SRD must be written in the language of the customer and the user as opposed to the more formal specification language employed in later systems engineering documents. This ensures that the document is readable by all stakeholders and adequately addresses all of the users' needs. It is critical that the SRD is comprehensive, readable and unambiguous. The SRD is not intended to be the functional baseline of the system and should not be written using a formal specification language. In many cases the difference between the SRD and the system specification is marked because the user is operationally focused and uses very little technical language. In other cases, such as our aircraft system, the stakeholders include air crew, ground crew, maintainers, and engineers, all of whom will use relatively technical language in their SRD statements. The difference is always that SRD statements are made in more natural language, which is generally much more easily read and understood than the language of formal specifications.

Example 2.1: SRD Content

In our aircraft development example, the SRD will contain the ACME Air's requirements in the language of the customer/users. For example, among the many requirements, there will be such statements as the following:

- *The aircraft is to be capable from operating from any Class X airport in the world.*

- *The aircraft is to provide "class-leading" comfort for passengers.*
- *The aircraft is to be capable of being turned around to its next flight within 30 minutes.*

The potential audience for the SRD includes everyone who has an interest in the delivered system, from the eventual users, operators, and support personnel to the systems engineers and designers responsible for designing, developing, and manufacturing the system, as well as the test personnel responsible for confirming that the system is fit for its intended purpose. Each audience set provides different views of the requirements suggested by the SRD. Endorsement of the SRD by each of the stakeholders maximizes the confidence of system designers that all of the fundamental user requirements have been captured.

Systems engineers are then responsible for the effort that translates the user requirements in the SRD into the more formal requirement statements contained in the system specification that forms the functional baseline of the system.

Figure 2.2 summarizes the process involved with the identification of stakeholder requirements and the development of the SRD.

2.2.2   Identify Stakeholders

The first step in developing the SRD (Figure 2.2, process C1.1), therefore, requires the identification of the stakeholders responsible for all aspects of the new system. These stakeholders are responsible for writing and agreeing on the SRD prior to it being endorsed. The systems engineer also calls upon these stakeholders during later activities to assist and clarify with the requirements analysis effort.

Example 2.2: Stakeholders for the Aircraft System

> *There are a number of stakeholders within ACME Air who provide a major source of information relating to the functional and performance requirements of the aircraft system. Corporately, the business owners of the airline have a set of more commercial requirements for the aircraft, such as fuel efficiency, passenger and freight capacity, and so on. Users, including air crew and ground crew—personnel who will eventually operate the aircraft—will have significant input to the functional and performance requirements. Marketing people dictate the features necessary in the aircraft system in order to make the aircraft marketable to passengers. Maintainers*

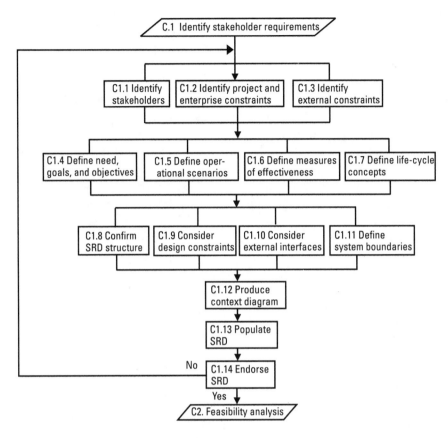

Figure 2.2 A summary of the process involved in the identification of stakeholder requirements.

influence the design to ensure that the aircraft is maintainable within the time and cost constraints set by the business case. Accessibility, reliability, and maintainability will be high on the priority list of the maintenance representatives. Logistics and supportability issues such as the provisioning of spare parts are also likely to influence design decisions.

2.2.3   Identify Project and Enterprise Constraints

Before focusing on the detail of the desired system, it is essential to identify the project and enterprise constraints (Figure 2.2, process C1.2) that are relevant to the system and its acquisition. This analysis provides essential information about the development environment for the system and begins the top-down approach to system development.

Enterprise constraints include any organizational policies, procedures, standards, or guidelines that guide system development and procurement. These constraints can include partnering relationships with other companies, contracting policies, and so on. For example, ACME Air may have a policy of only buying engines from a particular manufacturer so that through-life support issues are simplified across the entire fleet. This constraint must be articulated early as it has a significant effect on the procurement.

Project constraints include the resource allocations to the project as well as any externally imposed deliverables and acquisition time frames. Many companies have enterprisewide standards for processes such as quality assurance and systems engineering, and these methodologies guide the manner in which projects can operate. Additionally, the enterprise may require the project to report progress in a particular way or to implement particular metrics, tools, and documentation procedures.

2.2.4   Identify External Constraints

In addition to enterprise-imposed constraints, there are wider external constraints (Figure 2.2, process C1.3) on system development that arise from the requirement for conformance to national and international laws and regulations, compliance with industry wide standards, as well as ethical and legal considerations. Other external constraints include the requirement for interoperability and the capabilities required for interfacing to other systems. Again, an important aspect of top-down design is to understand these constraints before considering lower level system requirements.

2.2.5   Define Need, Goals, and Objectives

Because the user has most probably stated the need in a fairly general way, every project should begin with a concise statement of the need, elaborated by statements of the system-level goals and objectives (Figure 2.2, process C1.4).

The need statement should be quite short and may be expressed in only a few lines, although it should have a word or phrase for every important aspect of the system. While stakeholders often find it difficult to state the need in a single short sentence, the project is doomed to failure if the owners of the concept cannot describe it succinctly at the outset. For example, the need statement for our aircraft example might say something like "… to acquire a medium-sized aircraft that can provide class-leading comfort to passengers between Class X airfields on domestic and international routes."

The need statement is then expanded and qualified by short declarative statements of the system goals and objectives. Goals are normally relatively broad statements, each of which spawns a number of more specific objectives (although these are sometimes treated in the reverse order and objectives are considered to lead to goals). In our example, we would need to consider goals and objectives that elaborate on such matters as what is meant by "class-leading" and "medium-sized," how many passengers are to be carried, as well as operational issues, such as crewing and maintenance.

2.2.6 Define Operational Scenarios

Once the need, goals, and objectives have been articulated, the top-down process is continued through an examination of the range of operational scenarios (Figure 2.2, process C1.5) that the stakeholders propose for the system. The examination begins with a description of the general operational environment for the system to identify all of the environmental factors that may have an effect on the operation of the system. Specific operational scenarios are then described in users' language to depict the full range of circumstances under which the system is required to operate. It is not necessary to describe every possible scenario, but all types of operation must be represented. Scenarios also need to represent all stakeholder perspectives. For example, turning an aircraft around means different things to air crew and ground crew—both have valid views that spawn different sets of requirements.

These scenarios, or use cases, provide valuable guidance to the system designers and form the basis of major events in the acquisition phase, such as acceptance testing of the system as it is introduced into service. Despite any more detailed technical verification and validation procedures, the system's fitness for purpose is fundamentally related to its ability to perform in accordance with the operational scenarios defined at this stage.

In many cases it is also useful to define the various modes of operation for the system products under development. Designers need to understand if the system is to exist in a number of different modes, even if it is as simple as the difference between the fully operational mode or the training mode. Complex systems may have their requirements stated in a number of modes. For example, a modern fighter aircraft may have modes defined for air-to-air combat, ground attack, reconnaissance, naval operations, nontactical flights, and so on. Each mode must be associated with the particular conditions (mission, operational, environmental, configurational, and so on) that define it.

In our aircraft example, a number of modes may be defined for international and domestic operation including taxi, take-off, cruise, approach, landing, and turn-around. Modes may also be defined for maintenance and for administrative movement of the aircraft.

2.2.7 Define Measures of Effectiveness

An important top-level activity is the identification and definition of measures of effectiveness (MOEs)—Figure 2.2, process C1.6. An MOE is a metric by which the customer will measure satisfaction with products of the acquisition phase. Key MOEs include performance in each operational scenario, safety, reliability, supportability, maintainability, ease of use, and time and cost to train.

MOEs are supported by measures of performance (MOPs), which provide lower-level design requirements that are necessary to satisfy a measure of effectiveness. There are generally several MOPs for each MOE.

2.2.8 Define Life-Cycle Concepts

Early in the acquisition phase, the stakeholders must give some guidance on the life-cycle concepts (Figure 2.2, process C1.7) related to the development, production, testing, distribution, operation, support, training, and disposal of the system. While the systems engineering procedures that follow will ensure a life-cycle focus, it is important that the stakeholders focus on the major cost drivers that will have an impact on the supportability of the system. There are a number of life-cycle-related trade-offs at a business-case level. For example, the need might be for an inexpensive dashboard-mounted global positioning system (GPS) for a large trucking company. Because the item is to be procured in large numbers at a low cost, it may be deemed as part of the business case that it would not be cost effective to implement a repair system for defective items, which would instead be discarded when broken.

2.2.9 Confirm SRD Structure

Stakeholders must now agree on the structure and content of the SRD (Figure 2.2, process C1.8). The structure of Figure 2.2 provides a reasonable framework for an SRD, where headings for the relevant section are based on the boxes C1.1 to C1.7 and C1.9 to C1.11.

There are also a number of excellent references that provide guidance on process, content, and layout of the documentation. For example, ANSI/ AIAA G-043-1992 [2] suggests the following sections:

- *Scope.* This section provides an overview section describing in broad terms the system and its intended purpose.

- *Referenced documents.* It is critical to keep a record of the source of information in the SRD in case of conflicting and ambiguous requirements or some disagreement between stakeholders.

- *Operations.* This is a section in which the end user can record their requirements of the system, including the major mission(s), how these will be performed, techniques used, and any constraints.

- *Operational needs.* This section is derived from the previous section. Each mission is investigated to determine a series of requirements. The section also includes personnel requirements derived from the types of personnel involved in the system.

- *System overview.* This section includes a description of external interfaces, system states and modes, core capabilities, and any architectural considerations that need to be addressed.

- *Operational and support environments.* This section describes the conditions under which the system will be operated and maintained.

- *Operational scenarios.* A description of possible operational scenarios should be provided to help describe the likely events facing the system.

IEEE P1362 [3] also suggests that the SRD contains a description of the current situation and the need for a change. This helps put the system's intended purpose into context. P1362 suggests that the operational and organizational impact of the proposed system should be analyzed and discussed in the SRD. The consideration of alternative solutions can also be included into the SRD once feasibility analysis (see Section 2.3) is complete.

Once the content and structure is in place, a plan for populating, reviewing, and eventually endorsing the document should be agreed. Progress should be reviewed periodically throughout the development of the SRD.

2.2.10 Scoping the System

The next phase in successful requirements analysis requires an understanding of the scope of the system development effort (called *scoping*, in the vernacular). This scoping process helps to establish a clear understanding of what the system is expected to do by soliciting, determining, and detailing design constraints, external interfaces, and system boundaries.

Design constraints (Figure 2.2, process C1.9) include those factors that directly affect the way in which the system design can be conducted. Typical constraints include the state-of-the-art of relevant technologies as well as extant methodologies and tools to assist in the design, development, construction, and production of the system. Additionally, bounds such as all-up weight may be a design constraint for an aircraft system if it is to land on certain classes of airfield.

Interfaces with existing or future external systems (Figure 2.2, process C1.10) must also be defined, as these will place considerable requirements on the system under development. While these external systems are not directly related to the project, the success of the fielded system is often determined by its ability to interface to its external environment. For example, while it is possible to build a perfectly functional aircraft without consideration of air traffic control regulations, the aircraft would be useless because it would not be allowed to operate.

Definition of the system boundaries (Figure 2.2, process C1.11) is also critical to the success of the fielded system. It is essential that these boundaries are defined early in the acquisition phase so that it is clear which system elements are under the design control of the project and which are outside the control. This is also particularly important to the project manager, who is vitally concerned with defining what is to be included in the system as well as what is to be excluded.

To assist with the scoping process, a tool called a *context diagram* (Figure 2.2, process C1.12) may be used to illustrate the related systems, relevant regulatory environments, stakeholders, external systems, interfaces, and so on. Different systems may of course have significantly different context diagrams.

Example 2.3: Simple System Context Diagram for Our Aircraft Example

> *Figure 2.3 illustrates a simple context diagram for our aircraft example. In reality, of course, the diagram would be far more detailed.*

> Regulators. *Regulators include both domestic and international law makers who contribute to functional and performance requirements, although*

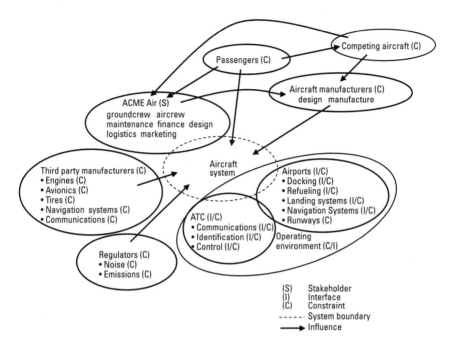

Figure 2.3 Simple context diagram for our aircraft example.

their input may be seen more as constraints on the system in terms of such factors as size, weight, and power restrictions, exhaust emission controls, noise limitations, and safety requirements.

Manufacturers. *Manufacturing processes impact on the design of the aircraft by placing constraints on aspects of the design in order to ensure producibility. Potential aircraft manufacturers need to be involved in the design process from early stages in order to influence the design decisions as required.*

Third-party manufacturers. *The aircraft industry contains a number of third-party companies who manufacture and sell aircraft components and systems, such as tires, avionics computers, and engines. Aircraft manufacturers make use of these off-the-shelf subsystems and components to minimize the costs and risks associated with aircraft design and development.*

This group of stakeholders contribute to the interface requirements for the aircraft system but may also contribute to functional and performance requirements.

Operating environment. *The operating environment is primarily defined by the aircraft system's mission and includes such things as airfields, airport facilities, weather conditions, temperature ranges, and time of day. The environment can often place some critical design constraints and requirements on the system and result in the requirement for a number of external interfaces.*

Passengers and competing aircraft. *Passengers and competing aircraft also provide constraints on the system in that if our aircraft is to provide class-leading comfort, the success of the acquisition will be measured by passenger satisfaction and the performance relative to other aircraft.*

2.2.11 Populate SRD

While there may be a large number of stakeholders who have an interest in the development of the SRD, the population of the SRD document (Figure 2.2, process C1.13) is generally best conducted by a small team of experienced customer representatives. It is important that all stakeholders are involved in the preparation of the document, however, and mechanisms such as structured workshops, surveys, interviews, and prototyping are necessary to ensure that all relevant information is considered.

2.2.12 SRD Endorsement

Endorsement of the SRD (Figure 2.2, process C1.14) is a critical step in the development of the system. The system specification cannot be developed until the SRD is complete and endorsed, as that document provides the foundation of the engineering effort that eventually results in the foundation of the functional baseline. While some may argue that the SRD and the more detailed engineering specifications can be developed simultaneously (with the consequent saving in time), the lower level specifications represent the results of analysis and synthesis and can only be completed adequately if these tasks are conducted on the complete set of input factors described by an endorsed SRD. An incomplete set of inputs results in an entirely inadequate analysis and synthesis process, which subsequently results in a substandard set of specifications. Parallel efforts also tend to result in confusion regarding the role, level of detail, and content of each document.

From a project management perspective it is also essential that the SRD is endorsed first so that the financial backers of the project can be assured that all stakeholders are unanimous in their agreement to the need, goals, and objectives of the proposed system and that all relevant input has been obtained. In light of the experience of previous project failures, it would be inappropriate to recommend the commitment of further funds to a project whose stakeholders have not formalized and endorsed the user requirements.

The SRD endorsement could occur in a formal meeting involving the key stakeholders or could be forwarded to key stakeholders for their individual review and sign-off.

2.2.13 Traceability

During the subsequent synthesis and analysis, systems engineers must ensure that each requirement in the SRD is traceable to at least one requirement in the system specification. In doing this, the customers can gain confidence that all SRD requirements have been addressed by the systems engineer, and they can gain an insight into the satisfaction of the SRD requirements. This is the start of a cornerstone systems engineering concept called traceability. Traceability from the SRD to system specification is called *forward traceability*. Systems engineers must also ensure *backward traceability* from system specification back to the SRD, which ensures that all requirements contained in the system specification are traceable back to at least one SRD requirement. This backward traceability ensures that additional requirements (not formally endorsed by the customer) have not crept into the system specification during the analysis and synthesis process during conceptual design.

Requirements creep must be guarded against as additional unendorsed requirements ultimately compete with endorsed requirements for valuable design space and may end up reducing the performance of the system against the endorsed requirements. Unendorsed requirements may also add significantly to the cost and schedule associated with the system development, without adding to the system's ability to solve the original stakeholder need.

2.3 System-Feasibility Analysis

Once the stakeholder requirements have been identified and articulated satisfactorily, the feasible system-level alternatives available to meet the new need can be developed. Feasible alternatives need to be considered in terms of available resources such as money, time, personnel, and materials. As mentioned in Section 2.2, some degree of feasibility analysis may be included in the SRD. The following steps need to be completed during feasibility analysis:

- Identify the possible system-level solutions capable of satisfying the need.

- Confirm that the requirements in the SRD are achievable and, if not, note the likely level of achievement.

- Evaluate the potential solutions in terms of feasibility, performance, effectiveness, technical and project risk, and other selected measures of performance.

- Recommend the best of the possible solutions (ensuring that the options are narrowed down as much as possible).

Figure 2.4 summarizes the process involved with system feasibility analysis.

A great deal of care should be taken during this stage of the project, as the feasibility analysis sets the direction for the rest of the system life cycle (remember the example of the RAAF F-111 in Section 1.2—the feasibility analysis could have an impact on a project for more than 70 years). To that end, life-cycle considerations such as reliability and maintainability should have a high priority during feasibility analysis.

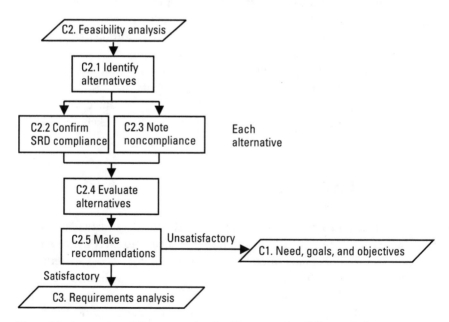

Figure 2.4 A summary of the process involved in system-feasibility analysis.

Example 2.4: Developing the Aircraft System

> ACME wants to service the short-to-medium domestic and international routes, but doesn't have the airframes to do it. The feasible alternatives are as follows:

> • Leasing or buying an existing aircraft;

> • Leasing or buying and modify an existing aircraft;

> • Contracting to have a new aircraft developed;

> • Outsourcing the operation to another airline;

> • Not servicing these medium routes.

There may be an opportunity to examine the marketplace to identify operational systems that may be assessed to assist with feasibility analysis. Assessing operational systems may also assist with the refinement of the need, goals, or objectives of the new system based on a realistic understanding of what is currently available. The feasibility of using existing systems to meet emerging needs may be assessed as a possible way of reducing the schedule and technical risk associated with the system development. The process is sometimes called *concept demonstration* or *technical demonstration*.

2.4 System-Requirements Analysis

The next step in conceptual design is requirements analysis, which, as the name suggests, concentrates on determining the system-level requirements of the system. The system-level requirements form the basis of the functional design of the system that is the main aim of conceptual design. System requirements analysis therefore represents the core of the functional design activities. Figure 2.5 summarizes the process involved with system requirements analysis.

The aim of system requirements analysis is to describe the requirements at the system level and be able to relate the functional design back to the set of endorsed need, goals, and objectives of the system contained in the SRD. Being able to relate the functional design back to the endorsed need establishes the traceability that is central and critical to systems engineering.

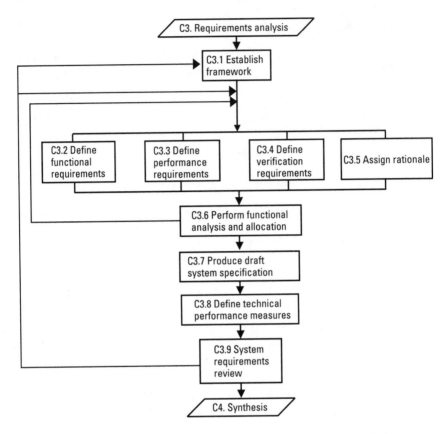

Figure 2.5 A summary of the process involved in system-requirements analysis.

During system requirements analysis, it is important to concentrate on what is required rather than how to do it. For example, a system-level performance requirement of the aircraft system might be a specified cruising speed. The physical subsystems that would be needed to perform this function (such as engines, airframes, and flight controls) may not be considered at this stage. The latter stages of the systems engineering processes are designed to focus on the physical subsystems and below and to determine how system-level requirements will be met through a disciplined design approach. Specifying how to meet the requirements at this embryonic stage may result in ill-informed design decisions leading to suboptimal designs. Because the customer is most likely to be performing the conceptual design tasks, specifying "how" will also have the undesirable effect of shifting some responsibility for achieving system performance from the contractor to the customer.

2.4.1 Establish Requirements Framework

The first step in the systems requirement analysis is to develop a requirements framework (Figure 2.5, process C3.1) around which the functional design of the system is to be based. Here we call the requirements framework the *requirements breakdown structure* (RBS). The words are deliberately chosen to differentiate this structure from the well-known project management document called the *work breakdown structure* (WBS). We demonstrate the relationship between the WBS and the RBS in Chapter 3, but at this stage it is sufficient to recognize that the two are not the same. The RBS is grouped by function, the WBS is structured by physical project elements and contains other project-related information.

The RBS for a system is the structure upon which the many system-level requirements are developed. The RBS framework provides a reference and a guide to those performing the analysis work during the development of the system functional baseline. In fact, the framework, when completely populated, can form the system's functional baseline. A representative requirements framework tailored to our aircraft system example is shown in Figure 2.6.

There are a number of advantages associated with the establishment of a suitable framework such as the RBS early in the requirements analysis activity. First, the framework will act as a reference source during requirements analysis to ensure that all aspects of the system requirements are addressed and that important areas are not omitted. The framework allows multiple people to work on the analysis simultaneously as it allows the effective allocation of responsibility for sections of requirements to individuals. The framework assists in avoiding duplication of requirements in different sections, which is undesirable as it raises the probability of conflict between requirements and often leads to ambiguity and confusion.

System requirements in the RBS must be complete and must fully describe the needs and constraints of the major stakeholders. The requirements should be objective, designable, measurable, demonstrable (or testable), and traceable. Although this sounds like an obvious statement, system requirements are often ambiguous and incomplete, conflict with other requirements, and are not traceable to any endorsed customer requirement. Customers sometimes prefer ambiguous and nonspecific system-level requirements in the mistaken belief that it does not lock them in to specific functions early in the project, or alternatively customers are not sure of their real requirements. Untraceable requirements sometimes appear in system specifications because the "needed" system documented in the SRD is not the same as the wanted

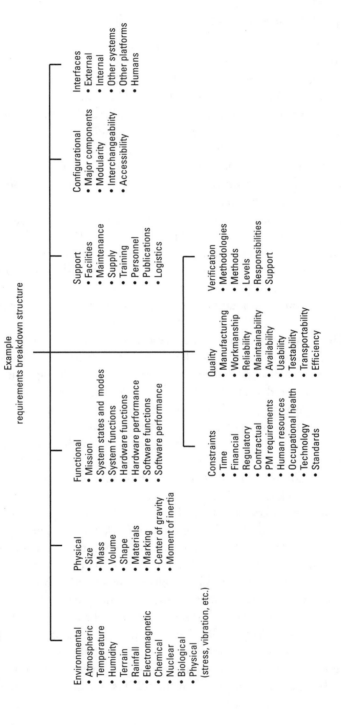

Figure 2.6 Example requirements-breakdown structure [4].

system that is described in the system specification. Regardless of why the system specification contains poor requirements, ambiguous and loose statements of requirement have an adverse effect on the remainder of the project and the system that is eventually produced. Time invested in getting the requirements correct significantly reduces technical, schedule, and financial risks associated with the system acquisition. The difficulties associated with the development of requirements are described in more detail in Section 7.1.5.

In performing the system requirements analysis, the following tasks are performed:

- Determining the major functions required of the new system (including maintenance, support, operational, and any other necessary functions);

- Assigning the necessary performance-related parameters or requirements to each of the functions as applicable;

- Articulating any other system-level requirements, such as the requirement categories shown in the example RBS;

- Determining the technical performance measures (TPMs) to be applied during the systems engineering effort.

Many tools exist to assist with the process of requirements analysis. For example, IEEE 1220 [5] is a systems engineering standard that suggests a procedure for performing requirements analysis. IEEE 1220 and other standards are introduced in Chapter 6.

2.4.2 Define Functional Requirements

Defining functional requirements (Figure 2.5, process C3.2) is an important part of the overall system requirements analysis that, as the name suggests, identifies the many different functions that the system needs to perform in order to meet the endorsed need, goals, and objectives contained in the SRD. The analysis identifies all functions, including maintenance functions, that are necessary to maintain or return the system to operational use. At this stage in the development, the concentration is still on what needs to be achieved, not how it should be done.

A possible way to analyze functional requirements is via FFBDs [6], a simple example of which is shown in Figure 2.7.

FFBDs are developed for the main purpose of structuring the system requirements into functional terms. The FFBD identifies the major system-

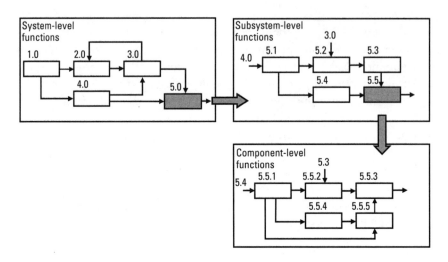

Figure 2.7 Example functional-flow block diagrams.

level (or top-level) functions that need to be performed by the system in accomplishing its mission. Each top-level block is expanded as necessary to ensure that the functionality of the system is adequately defined. This may require the functional analysis to break the system down further than the major system-level functions even at this early stage. The requirements should be "decomposed" as far down as necessary to achieve this aim. The FFBD helps the designers evaluate each function adequately and assists in the definitions of the requirements in terms of performance, inputs, outputs, controls, and constraints, which is the next stage in conceptual design.

Once the FFBD has been completed to the required level, all major system functions and their interrelationships should have been identified. Functional allocation may occur where the functions are grouped together in logical groups and assigned to one of the headings in the RBS. This helps when system-level performance requirements are considered but also ensures that the system specification that results from the population of the RBS is logical and readable and avoids duplication and potential conflict. This forms a sound framework within which preliminary design can be conducted.

Note that the FFBD shown in Figure 2.7 makes use of a numbering system. The system-level functions have been assigned numbers, and those numbers have been continued through the lower level functions. In this way, even the lowest level functions and requirements can be traced back to the original system-level function. Traceability has already been introduced as a cornerstone in systems engineering, and it becomes increasingly important as the design matures and becomes more complex.

Operational requirements are normally given the highest priority during the early stages of a project because operational requirements tend to be the most tangible and most easily understood. The life-cycle approach emphasized in systems engineering also acknowledges that design decisions made early in the acquisition phase can have a large impact on the maintenance and support of the system. To that end, maintenance and support requirements need to be determined and specified during conceptual design. In this way, the design of the system ensures efficient and effective support during the utilization phase. Human factor considerations also produce a number of additional requirements. Functional requirements for the system are obtained from an analysis of design space limitations, climatic limits, eye movement, operator reach, occupational health and safety issues, ergonomics, cognitive limits, usability, and so on.

2.4.3 Define Performance Requirements

Once the functions have been identified and grouped according to the agreed RBS, the systems engineers and stakeholders must agree on the performance-related parameters (Figure 2.5, process C3.3) that the new system must achieve. Having decided what the system must do, the designer must now determine how well the system is to perform each of those functional requirements. A good discipline is to ensure that every time a functional requirement is articulated, a corresponding performance statement is made.

Most of the operational functions will have obvious performance parameters associated with them, such as speed, accuracy, endurance, and acceleration. Support and other functions also require parametric definition to define completely the requirement. The following is a nonexhaustive list of examples of these parametric requirements:

- Environmental considerations, including temperature ranges for both operation and support, humidity tolerances, and operation under conditions of obscuration, such as rain, snow, or fog (operation under mechanical stresses such as noise, vibration, and shock should also be considered, as should chemical, nuclear, and biological situations, if applicable);

- Quality factors, such as availability requirements, reliability and maintainability limits, and manufacturing tolerances;

- Support requirements, including storage capacity, spares allowances, personnel establishments, training goals, and maintenance targets (the intended levels of maintenance employed to support the system once it becomes operational should also be included and related to the overall logistics support concept described or referenced in the SRD);

- Utilization requirements, including duty cycles, hours of operation per day, and operational days per year.

2.4.4   Define Verification Requirements

The definition of functional and performance requirements is not complete until verification requirements (Figure 2.5, process C3.4) are also included. It is good practice to ensure that every time a functional requirement is articulated, a corresponding verification statement is made. It is often difficult to write verification requirements for system-level functions, but the discipline of doing so is important because there is little point in stating a requirement for a function without consideration of how the function is to be tested. An additional benefit of adhering to the discipline at this stage is that test plans become much easier to write because the tests required at any level can be considered within a framework provided by an aggregation of the verification requirements at that level.

2.4.5   Assign Rationale

It is also useful to record the rationale (Figure 2.5, process C3.5) behind each requirement, which will further assist with the removal of ambiguity but more importantly will assist in situations where more than one person is involved in the requirements analysis over the course of the project. The rationale explains why each requirement is necessary and the logic behind performance imperatives assigned to the requirement.

An example commonly used in systems engineering circles involves the design of a back windscreen of a passenger car, which had a user requirement to be able to withstand the air pressure associated with the car moving at 80 km/h. During the design it was established that the car's maximum speed in reverse gear would only be 40 km/h, so designers reduced the requirement for the rear windshield to withstand air pressure from 80 km/h to 40 km/h. Such a decision seemed to make sense, particularly because there was likely to be a subsequent cost saving in the back windscreen component. When the new vehicles were reversed onto a car transporter for delivery from the factory to retail outlets, however, the rationale for the original user requirement became readily apparent.

2.4.6 Perform Functional Analysis and Allocation

Once functional, performance, and verification requirements have been defined at an upper level, designers begin to analyze and allocate those functions. The aim of functional analysis (Figure 2.5, process C3.6) is to ensure that the elements of a system-level functional architecture are defined at sufficient depth to allow for synthesis of solutions in the following stage.

The requirements articulated by the user (gathered in the previous stages Figure 2.5, process C3.2 to C3.5) are generally at a high level and often tend to be more qualitative than quantitative. In fact, the major requirements collected in the first pass will look very similar to the key statements of the SRD. For example, our stakeholders require that the new aircraft is to be able to operate from any Class X airport. While stakeholders will articulate some additional requirements during requirements elicitation/generation (see Section 7.1.2.1), these broader functions need to be analyzed to identify the lower level requirements necessary to achieve the parent requirement. These lower level requirements are often called *derived* requirements because they are not stated directly by a stakeholder, but are an artifact of the requirements-analysis process. For example, if the parent function is to be able to operate from any Class X airport, designers would derive requirements related to minimum runway lengths, runway surfaces, maximum allowable aircraft weights, and so on.

Example 2.5: System-Level Requirements

> *In Section 2.2.1, we briefly discussed some potential SRD statements made by stakeholders in the aircraft system. Five possible SRD statements relating to our aircraft example are listed next, along with examples of derived system-level requirements. Note that a numbering convention has been added to assist with traceability during later design activities.*

> 1. *Operate from any Class X airport*
>   1.1 *Minimum runway lengths*
>   1.2 *Runway surfaces*
>   1.3 *Maximum allowable aircraft weight*
>   1.4 *Essential navigation aids*
>   1.5 *Essential automatic landing systems*

    1.6 Essential communications systems

2. *Provide class-leading comfort to passengers*

    2.1 Minimum amount of leg room
    2.2 Minimum seat dimensions
    2.3 Minimum seat support requirements
    2.4 Entertainment systems
    2.5 Bathroom facilities
    2.6 Catering services

3. *Ability to turn the aircraft around in less than 30 minutes*

    3.1 Conduct refuelling operations
    3.2 Loading requirements for passengers, cargo, and catering
    3.3 Unloading requirements for passengers, cargo, and catering
    3.4 Operational maintainability requirements

4. *Be a commercially viable aircraft system*

    4.1 Minimum aircraft range
    4.2 Minimum cruising speed
    4.3 Fuel economy requirements
    4.4 Minimum capacity (passenger/cargo)
    4.5 Maximum numbers air crew/cabin crew/ground crew required

5. *Conform to all relevant regulatory requirements*

    5.1 Rates of climb/descent
    5.2 Minimum levels of safety equipment
    5.3 Minimum numbers of engines for over water operations
    5.4 Flight recording capabilities
    5.5 Emergency egress requirements

Naturally, there are many thousands of requirements needed to fully define the functional design of the aircraft system. We obviously do not have space here to include them all and only list those sufficient to illustrate the concepts associated with conceptual design and preliminary design. As mentioned, the numbering system has been provided to support traceability throughout the process. For example, the requirement 3.1 Conduct Refuelling Operations relates directly to SRD statement 3.0 Ability to turn the aircraft around in less than 30 minutes.

FFBDs are very useful tools at this stage, both as analysis tools and as convenient mechanisms for communication with stakeholders.

Example 2.6: Use of FFBDs During Conceptual Design

During conceptual design, designers use FFBDs to address system-level functions. An FFBD describing the turnaround operation of the aircraft may take the form shown in Figure 2.8. Turnaround involves processing the passengers, luggage, freight, and so on between flights and preparing the aircraft for its next flight. Note that many such FFBDs are required to cover the other functional categories required of the aircraft system.

The SRD requirement 3.0 to turn the aircraft around resulted in a number of system-level requirements, including 3.1 Conduct refueling. We can see that 3.1 Conduct refueling must take no more than 30 minutes to complete. This is an example of a functional requirement (refueling) and an associated performance parameter (less than or equal to 30 minutes).

Note that, even at the highest level, the FFBD in Figure 2.8 contains requirements that may not have been directly articulated by the stakeholders in the SRD. For example, the functional flow for this series of activities begins with landing the aircraft, a requirement that is implied in the user requirement to turn the aircraft around in 30 minutes, but may not have been stated explicitly. The designers are responsible for ensuring that these implied requirements are captured early on in the process and do not lie dormant until identified during user evaluations later in the process.

Functional analysis also involves investigation of the relationships between requirements. In addition to the derivation of requirements during

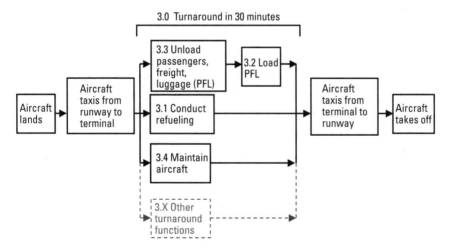

Figure 2.8 An FFBD describing aircraft turnaround.

functional analysis and allocation, system designers also *group* and *allocate* functions.

The RBS is an ideal tool with which to collect and group requirements.

Example 2.7: Aircraft System-Grouping and Allocation of Requirements

A possible RBS for our aircraft example is shown in Figure 2.9 to illustrate the process of grouping and allocation. Of course in a real project, there will be many more requirements than those shown in Figure 2.9.

Note in Figure 2.9 that the requirements now have two numbers associated with them. The RBS structure requires that each requirement is allocated an RBS number in accordance with the RBS hierarchy. Each of these requirements comes from the SRD, however, and already has an SRD number. For each RBS requirement with an RBS number, there is now a number in brackets to show the appropriate SRD number. This dual numbering system shows the relationship, or traceability, between each RBS requirement and the SRD. It is critical that these numbers or relationships are maintained to facilitate traceability up and down the hierarchy. Both numbers are effectively an attribute of each individual requirement.

Allocation has two aspects. First, functions are allocated to the appropriate group of the RBS. Second, higher level requirements flow down to lower level requirements. For example, the aircraft must be able to be turned around within 30 minutes. Figure 2.8 shows that this requirement consists of the lower level functions of unload passengers, conduct refueling, and maintain aircraft, which can all be conducted in parallel within the time frame.

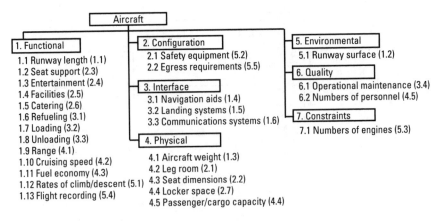

Figure 2.9 An example RBS after grouping and allocation.

We can therefore allocate the performance requirement to lower functions and can allocate the performance measure so that we state that unloading passengers, refueling, and maintenance must each be completed within 30 minutes.

Figure 2.5 shows that there is a feedback loop between functional analysis and the definition of functions and their rationale. This is an important iterative loop, at the end of which designers should be confident that they have captured all relevant system-level requirements at a sufficient level to support synthesis.

2.4.7 Produce Draft System Specification

The complete and populated RBS forms the basis for the system specification, which, when approved, becomes the system's functional baseline. At this stage, however, the populated RBS forms the basis for the draft system specification (Figure 2.5, process C3.7) because the requirements collected and analyzed so far have yet to be synthesized into an architectural solution. There will therefore be some degree of latitude allowed in certain requirements and labels, such as mandatory, important, and desirable, which leave some flexibility for subsequent design.

2.4.8 Define TPMs

The parametric requirements determined during the preceding stages assist in developing quantitative criteria (Figure 2.5, process C3.8) called TPMs that the system must meet. TPMs are identified early in the system development effort and are continually monitored and tracked throughout the system development as a means of managing the risks associated with system development.

The first step in identifying TPMs is to identify the quantitative parameters that require tracking throughout the project. Once the parameters have been identified, they should be prioritized in terms of their importance as viewed by the customer. A second list of TPMs may be established, prioritized and maintained by the contractor. As is often the case with subjective issues such as risk, a second perspective is always valuable. Both lists will imply priority to potential suppliers of the system during its design and development. This level of priority will assist the designer in determining the design emphasis to be placed on achieving each of the TPMs and the potential for trading off the performance of one TPM against the performance of another.

Throughout the design process, a list of TPMs and associated metrics should be maintained along with the priority, the benchmark objective, and the current level of achievement and projected/estimated performance. It is normal for some TPMs to move down in the priority order or drop from the list entirely as the design proceeds because their ultimate achievement is no longer uncertain. It is also normal for activities that were not considered risky during early stages of design to be added or moved up in priority due to previously unforeseen complications. This assists both the customer and the supplier in the identification and mitigation of technical risks during the acquisition phases. IEEE 1220 [7] states that appropriately selected TPMs can be used to assess conformance to requirements, assess conformance to levels of technical risk, trigger development of recovery plans for identified deficiencies, and examine marginal cost benefits of performance in excess of requirements.

Bands of acceptable variation from the expected value will also be determined and agreed upon so that unnecessary risk management actions are not instigated at the first sign of a very minor variation. The variation bands will be typically large in the early stages of the design but will become increasingly narrow as the design matures.

An example may clarify the role of TPMs in the acquisition phase of the systems engineering effort and emphasize the uses of TPMs as described by IEEE 1220.

Example 2.8: TPMs

Our aircraft system must be capable of a minimum rate of climb in order to comply with regulatory requirements governing departures from airfields. The designers of the aircraft system may identify rate of climb as a major design challenge and therefore a candidate for tracking using the TPM process. Designers can estimate the likely rate of climb of an aircraft system of this type, but their estimate will not be confirmed until the design of the aircraft matures.

In the early stages of a system life cycle, it is reasonable to expect that the aircraft design will exist on paper and in computer models, allowing an initial estimate of rate of climb to be made. The design will continue as the subsystems are designed and incorporated into the overall airframe design. Each subsystem will add weight to the aircraft and may change the aerodynamic performance of the aircraft (for example, external antenna systems). To that end, the subsystem design will start to alter the rate of climb that the aircraft is likely to achieve, thereby refining the initial estimate. During the latter stages of design, a prototype aircraft may be produced and test flown. Among other things, this test flying will confirm the validity of the earlier estimates of rate of climb and establish actual rates of climb achievable under different circumstances.

To monitor the risks associated with this critical TPM, the aircraft rate of climb could be estimated, investigated, and reported upon at discrete stages throughout the acquisition phase. Design reviews are a convenient forum for such activities. The progress of rate of climb towards the required rate of climb is of interest to both the designers of the system and the customers of the system. Failure of the aircraft system to meet this requirement will prevent the aircraft from meeting all of its original (and endorsed) needs, goals, and objectives (in this case, the requirement to operate from Class X airfields).

The graphs in Figures 2.10 and 2.11 show two possible profiles of rate of climb when 50% of the system design is complete. Although both profiles show the same rate of climb estimate at the 50% mark, the likely outcome from each scenario is very different.

In Figure 2.10, the rate of climb is decreasing as additional subsystems are designed and added to the aircraft. The decrease is staying well within the tolerance band and is tending toward the minimum acceptable rate of climb in a linear (predictable) fashion. At the 50% mark, the aircraft system has an estimated rate of climb of x feet per second. In this scenario, the rate-of-climb TPM would be presented at the relevant design review, and it is likely that the reviewers would not see any real problems with rate-of-climb TPM.

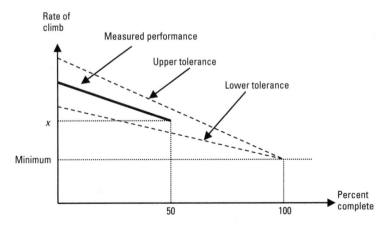

Figure 2.10 Rate-of-climb TPM as a function of percentage of work complete—Example 1.

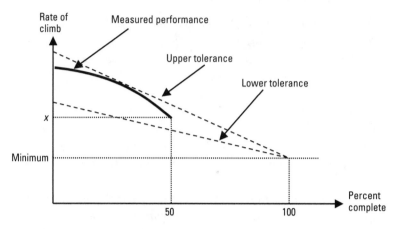

Figure 2.11   Rate-of-climb TPM as a function of percentage of work complete—Example 2.

In Figure 2.11, the rate of climb estimate at the 50% complete mark is also x feet per second. If the instantaneous estimate were the only indicator of rate of climb, it would seem reasonable enough. However, when the trend is observed, the rate of climb estimate brushes the upper tolerance band and then starts to dip alarmingly towards the lower tolerance band. Presented with this scenario, reviewers should note the trend towards the lower tolerance band and raise some concerns. In all likelihood, there may be a reasonable explanation for the trend. For example, the heaviest subsystems may have been designed during the first 50% of the project and the remainder of the subsystems are not expected to detract too much further from rate of climb. However, if there is no reasonable explanation, more detailed investigation is required to support risk management.

TPM tracking, as demonstrated in this example, should occur throughout the acquisition phase, and the parameters should be assessed periodically during technical reviews. TPM management (including identification, prioritisation, and setting of values) is therefore an important part of conceptual design.

2.4.9   System-Requirements Reviews

System requirements reviews (SRRs) may be conducted periodically throughout conceptual design to verify and approve sets of system-level requirements. The number of SRRs depends on the size and complexity of the system. Large and complex systems may require multiple SRRs during conceptual design whereas simple systems may not require any. It is recommended that at least one SRR is held, however, even in simple projects, so that the stakeholders have some formal involvement in the transition from the SRD to the system specification.

The aim of SRRs is to progressively monitor and approve the system-level requirements that are developed on the way to the functional baseline. Progressive reviews allow the requirements analysis effort to continue with confidence in the preceding decisions. SRRs may or may not be considered formal reviews. SRRs also ensure a complete understanding of the requirement being proposed prior to the establishment of the functional baseline.

Requirements reviewed at this stage are likely to contain margins of acceptable performance and may be categorized according to priority. A possible categorization may see requirements listed as mandatory, important, or desirable where mandatory requirements are absolutely essential, desirable requirements are nice to have, and important requirements are somewhere in between. SRR will review the margins and category of each requirement to ensure that each is being afforded appropriate weighting.

Other information such as manufacturing plans, design schedules, and personnel requirements plans is also reviewed at the SRR. Once the SRR has been completed, the design activities associated with conceptual design get underway in earnest.

2.4.10 Other System-Level Considerations

Functional, performance, and verification requirements are captured in the RBS, which, when complete, forms the structure for the system specification. However, a number of additional considerations also need to be captured. These additional issues relate to how the customer expects the system development to proceed and to management issues such as project management, systems engineering, and logistics.

The following list provides a nonexhaustive set of example issues that will need to be considered and specified prior to the commencement of system development. As described, most of these examples will not reside in the RBS (system specification), but may be more at home in the associated contractual documentation, such as a statement of work (SOW):

- Systems engineering requirements, including test and evaluation expectations, configuration management requirements, documentation deliverables, technical reviews and audits, and adherence to certain engineering standards (covered in much more detail in

Chapter 5). These requirements need to be considered and included because they will dictate how the design and development effort will be organized and managed. Systems engineering requirements add expense and time to the project but should reduce the technical risks of the system development effort.

- Logistics support concepts including maintenance levels, repair policies, and organizational responsibilities. All traditional aspects of logistics, including spares, transportation, warehousing, and data should be considered and included.

- Organizational responsibilities for the project management, systems engineering, logistics management, operation, and support should be determined and clearly articulated so that all stakeholder organizations are well aware of their respective responsibilities during the different stages of the system's life cycle. Some of these requirements are contained in the contractual documentation, and some are contained in internal customer processes and procedures.

2.5   System-Level Synthesis

The system design has now progressed to the stage where some of the system-level design decisions can be made. System requirements analysis has identified functional, performance, and verification requirements and has collected the functions into logical groups that now require satisfying via some design decisions. Synthesis takes these alternatives and develops the preliminary concepts and establishes the relationships between system elements. Synthesis (at the conceptual design level) establishes a system configuration that is representative of the final system form. The configuration established at this stage is not assumed to be final as the design is very immature and may go through significant changes later on in the design process.

Trade-off analyses should be conducted to ensure that the design is the best alternative available. Major decisions such as potential use of commercial off-the-shelf (COTS) products need to be made early in the process. Trade-off analysis is another name for a decision-making process. The alternatives are identified, the criteria set, and the most appropriate alternative is selected. The process of conducting a trade-off analysis is detailed in Chapter 7. Figure 2.12 summarizes the process involved with system-level synthesis.

Based on the results of the requirements engineering and analysis, a range of architectural options are developed (Figure 2.12, process C4.1) that provide potential system solutions. The selection of one of these options as

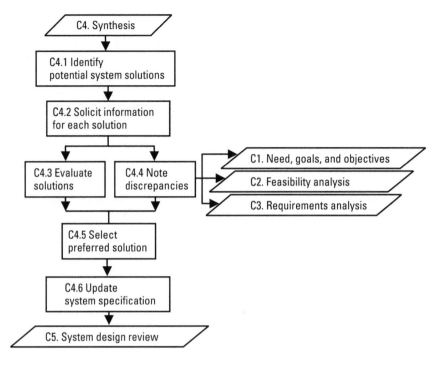

Figure 2.12 A summary of the process involved in system-level synthesis.

the preferred solution requires information to be collected (Figure 2.12, process C4.2) from the systems engineering process, life-cycle costings, quality assurance, test and evaluation, maintenance, integrated logistics support, and so on. The potential solutions are then evaluated (Figure 2.12, process C4.3), which requires the development of suitable evaluation criteria as well as an evaluation framework. Any discrepancies in the systems engineering products are noted (Figure 2.12, process C4.4) and fed back to previous activities. The preferred solution is then chosen (Figure 2.12, process C4.5).

In some applications of the systems engineering process, the system-level synthesis may take the form of responses to a request for tender (RFT). In this case, potential contractors are provided with a draft system specification upon which system-level synthesis is conducted. The customer then evaluates the tender responses (which represent feasible system-level designs) before selecting the preferred system-level solution. In the process, the draft system specification is further refined in an attempt to make it more suitable as the system's functional baseline and to take into account the range of feasible alternatives. As feasible alternatives are presented, the customer may need to revisit the SRD and seek endorsement from stakeholders for proposed changes. Some of the endorsed requirements may not be feasible within the approved schedule or cost constraints. This process is the first test of traceability as these discrepancies are noted and addressed by the stakeholders.

The final stage in conceptual design is the updating of the system specification (Figure 2.12, process C4.6), which documents the system functional baseline and includes the results from the needs analysis, feasibility analysis, system requirements analysis, and the critical TPMs.

The draft system specification produced at the end of requirements analysis necessarily contains margins of acceptable performance and requirement priorities. Potential system solutions have now been investigated (at least in broad terms) and the preferred approach to the problem has been determined. With this approach in mind, the draft specification needs to be refined to remove the ranges of acceptable performance and to reflect exactly what is to be achieved by the system and the minimum acceptable level of performance. The refined system specification and a broad system solution for achieving the specification requirements are the main products of synthesis.

The system specification is perhaps the most important of all systems engineering documents because it becomes the source of reference for all of the subordinate specifications that are produced at later stages through the design process. If the system specification is solid, it forms an effective foundation for the remainder of the design and development effort. Errors or omissions in the system specification flow into the remaining design effort. The later these errors are discovered, the more expensive and time consuming the rectification becomes.

The system specification must have its roots in the SRD, as the SRD is the source of endorsed stakeholder requirements. Systems engineers are responsible for expanding upon SRD requirements and producing a set of detailed requirements in the system specification that, when combined, ensure that this set of SRD requirements are met. Although the system specification is drawn from the SRD, it is much more detailed and has translated the SRD statements in the users' language into a more formal specification language.

Example 2.9: SRD Versus System-Specification Language

The language utilized in the SRD is intended for easy reading for a larger audience than the specification language, which tends to be more precise. For example, the SRD requirement statement, "The aircraft is to be capable of operating from any Class X airport in the world," will spawn, among many other requirements, the more detailed system specification statement, "The aircraft undercarriage shall exert no more than y kN/m^2 *of pressure on the tarmac when the aircraft is stationary and at maximum all-up weight."*

The system specification may take any one of many forms [8], and the most appropriate form depends on the application. It is safe to assume, however, that the system specification should contain at least the information described in the completed RBS.

The system specification is sometimes called a type A specification. The lower level specifications (types B to E) and their relationship with the system specification are discussed in detail in Chapter 5. While the format of the system specification varies between organizations, a suggested format for a paper-based specification is given in Section 7.1.

2.6 System-Design Review

It is normal to conduct some form of review at the conclusion of each of the design phases, as this forms part of the evaluation function. The system design review (SDR) achieves a number of goals, including a formalized check of the conceptual design with respect to the specified requirements; a formalized communication of the intended design approach to the major players in the design effort; a means by which interface issues can be raised, discussed, and resolved; and a formal record of design decisions and acceptance. The SDR is a systems engineering management function and is detailed in Chapter 5.

Figure 2.13 summarizes the process involved with SDR. SDR occurs once the entire set of requirements has been placed in the system specification. It is essential in all reviews to establish the entry and exit criteria (Figure 2.13, process C5.1) for the review—that is, it must be clear what must be in place before the review begins and what must have been achieved before the review can be considered to be complete. Additionally, much time will be wasted unless all of the necessary documentation is prepared and reviewed (Figure 2.13, process C5.2) in advance of the review, and all parties have agreed to the agenda for the review (Figure 2.13, process C5.3).

The review then looks at each requirement and ensures that the requirements are met by the preliminary system architecture determined during system-level synthesis (Figure 2.13, process C5.4). If all is satisfactory in these regards, the system specification is approved and the functional

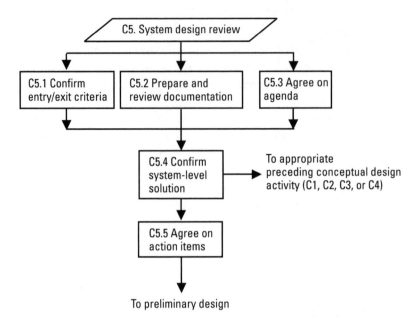

Figure 2.13 A summary of the process involved in the SDR.

baseline is established. It is rare that SDR ends with all outstanding issues resolved during the review because many will take some time to resolve. Rather than hold up subsequent design, a number of action items are agreed upon (Figure 2.13, process C5.5) to account for outstanding action from the review. These actions are completed in parallel with the early preliminary design activities and are reviewed for completeness in conjunction with a later review or audit.

The delineation between conceptual design phase and preliminary design is often not clear. SDR aims to provide some delineation by performing the following functions:

- *Evaluation of the system-level design proposal.* This evaluation investigates the activities of conceptual design in an attempt to estimate the ability of the proposed system-level designs to meet the system-level requirements (remembering that these requirements include operational requirements and support requirements, while the system design includes hardware, software, personnel, facilities, and so on). This may involve a review of the major subsystems and their role in achieving system performance.

- *Approval of the system specification.* The system specification is the document that needs to reflect the system-level requirements approved during SRR. New system-level requirements may need to be added to the system specification and errors or conflicts in existing requirements need to be clarified prior to the final approval of the system specification. It is likely that the acceptable margins determined during SRR will be replaced by a more acceptable performance level for each requirement. It is also likely that the categories (mandatory, important, and desirable) will be merged so that each statement in the specification becomes an unambiguous requirement.

- *Establishment of the initial functional baseline.* Through the approval of the final system specification, the initial functional baseline for the system is established. This is the baseline from which the remainder of the design will follow.

Endnotes

[1] ANSI/AIAA G-043-1992, *Guide for the Preparation of Operational Concept Descriptions*, American National Standards Institute, American Institute of Aeronautics and Astronautics (sponsor), 1993.

[2] ANSI/AIAA G-043-1992, *Guide for the Preparation of Operational Concept Descriptions*, American National Standards Institute, American Institute of Aeronautics and Astronautics (sponsor), 1993.

[3] IEEE P1362, *IEEE Guide for Information Technology-System Definition-Concept of Operations Document*, New York: IEEE Software Engineering Standards Committee, 1998.

[4] Based on the content of Figure 3 in ECSS-E-10A, *Space Engineering-System Engineering*, Noordwijk, the Netherlands: European Cooperation for Space Standardization, 1996.

[5] IEEE-STD-1220-1998, *IEEE Standard for Application and Management of the Systems Engineering Process*, New York: IEEE Computer Society, 1998, p. 61.

[6] In B. Blanchard, and W. Fabrycky, *Systems Engineering and Analysis*, Upper Saddle River, NJ: Prentice-Hall, 1998. Blanchard and Fabrycky note that these generic FFBDs can be prepared by any one of the graphical methods including integrated definition modeling method, the behavioral diagram method, and the N2 charting method, which are compared and discussed in B. Blanchard, D. Verna, and E. Peterson, *Maintainability: A Key to Effective Serviceability and Maintenance Management*, New York: John Wiley & Sons, 1995. The Defense Systems Management College

management guide (*Systems Engineering Management Guide*, Washington, D.C.: U.S. Government Printing Office, 1990, pp. 6.3–6.5) explains how FFBDs can be used during function identification.

[7]   IEEE-STD-1220-1998, *IEEE Standard for Application and Management of the Systems Engineering Process*, New York: IEEE Computer Society, 1998, p. 61.

[8]   Sources of information on the content of the system specification can be found in the following:

Data Item Description, System/Subsystem Specification, DI-IPSC-81431, DD Form 1664, U.S. Department of Defense, April 1989.

ECSS-E-10A, *Space Engineering-System Engineering,* Noordwijk, the Netherlands: European Cooperation for Space Standardization, 1996.

MIL-STD-961D, *Department of Defense Standard Practice for Defense Specifications,* Washington, D.C.: U.S. Department of Defense, 1995.

MIL-STD-490A, *Military Standard—Specification Practices,* Washington, D.C.: U.S. Department of Defense, 1985.

3

Preliminary Design

3.1 Introduction

Preliminary design starts with the functional baseline defined during conceptual design and continues with the effort to translate system-level functional requirements into design requirements for the subsystems that will be combined to form the system. This translation requires a continuation of the requirements analysis started in conceptual design. Specific requirements for the hardware, software, and personnel making up the system need to be determined. Trade-off studies are conducted, and the result of the preliminary design effort is the establishment of an *allocated baseline*, where requirements are allocated to specific subsystems making up the system.

Responsibility for performing preliminary design normally rests with the contractor, who accepts responsibility for the system meeting the requirements of the functional baseline (normally prepared by the customer). The customer's role now increasingly becomes one of monitoring, reviewing, and supporting contractor progress. The customer normally avoids becoming actively involved in design decisions made during preliminary design (or any subsequent activity, for that matter), as the responsibility for the resultant functionality and performance of the system rests with the contractor.

In many cases, the customer may opt to add an additional level of rigor to the process by engaging independent systems engineering consultants to provide independent review of the many and varied engineering documents and products produced by contractors during the course of a systems development.

The activities conducted during the preliminary design effort include subsystem requirements analysis, subsystem requirements allocation, subsystem synthesis and evaluation, and preliminary design review (PDR). These tasks are illustrated in Figure 3.1 and described in the following sections.

3.2   Subsystem-Requirements Analysis

Requirements analysis has already been described as the process through which requirements are progressively decomposed from the system level to the subsystem level and so on until all functions, parameters, and interfaces have been defined and the necessary resources have been identified to meet the original user requirements. The major activities in requirements analysis are described in Figure 3.2.

Requirements analysis in preliminary design is aimed at continuing the effort commenced during conceptual design to the next level of detail. Because preliminary design will most probably be conducted by a different team than that which conducted conceptual design (that is, by the contractor rather than the customer), the subsystem requirements analysis activity begins with a review of the system specification (Figure 3.2, process P1.1) and a review of the relevant TPMs (Figure 3.2, process P1.2).

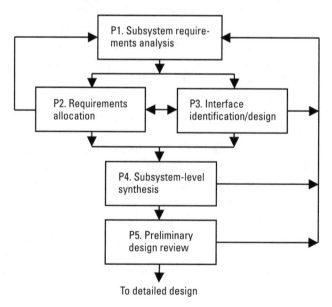

Figure 3.1 Preliminary design activities.

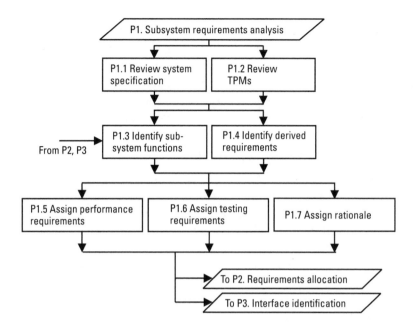

Figure 3.2 Requirements-analysis activities.

The FFBD is a relatively intuitive concept that was introduced in Chapter 2 as a means of performing an analysis of the functions required of the system. The diagrams that are generated during the preliminary design effort are more detailed than in conceptual design and are aimed at identifying (Figure 3.2, process P1.3) and deriving (Figure 3.2, process P1.4) more detailed functional requirements at the subsystem level. FFBDs also assist in defining the sequences and interrelationships between the blocks (serial versus parallel and so on) as well as the major functional interfaces. Performance (Figure 3.2, process P1.5) and testing (Figure 3.2, process P1.6) requirements are then assigned, as is a rationale (Figure 3.2, process P1.7) for each functional requirement.

It is not possible to prescribe the level of detail that must be derived during subsystem requirements analysis. The detail must be sufficient to completely define the major subsystems comprising the system. The detail produced during preliminary design is used either to procure or to design and produce the major subsystems and their components during subsequent stages of the systems engineering process.

Each block of the FFBD is eventually considered individually in determining the necessary inputs, the expected outputs, the constraints placed on the function, and finally the mechanisms by which the function could be achieved. Note that this stage is the beginning of the conversion of what the system needs to do into how it is going to do it, which is more commonly referred to as the translation of functional design into physical design. Sometimes, as described in Section 3.3, a number of different functions may be performed by the same piece of equipment or piece of software and so on.

Example 3.1: Detailed Aircraft System FFBD

> *This example illustrates how an FFBD may be used during preliminary design to define system functionality at varying levels of detail.*
>
> *In Section 2.4, we suggested that an FFBD describing the actual turnaround operation of the aircraft may take the form shown in Figure 3.3 (a copy of Figure 2.8). The SRD requirement 3.0 to turn the aircraft around resulted in a number of system-level requirements, including 3.1 Conduct refueling.*
>
> *Let's investigate this function further as we would during preliminary design. By using the FFBD process to investigate the requirement further in Figure 3.4, we can see that refueling an aircraft involves a number of steps prior to and after the actual refueling process. We can also see that once we*

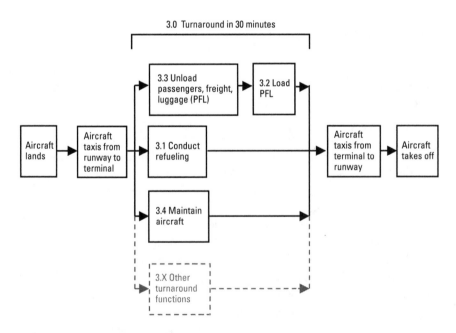

Figure 3.3 An FFBD describing aircraft turnaround.

have taken the other activities into account, the actual refueling process (3.1.6 Refuel aircraft) must now take place inside 17 minutes. In the worst case, therefore, we must be able to refuel the aircraft from a completely empty state to maximum fuel capacity in less than or equal to 17 minutes.

The requirement to refuel in less than or equal to 17 minutes has been derived using the FFBD process and is an example of a derived requirement that would be placed in our RBS under the 1. Functional heading and the 1.6 Refueling subheading. It is also an example of allocation of the upper level performance requirement of 30 minutes down to 17 minutes for the refueling process. We do not have space here, but in a real project we would continue the FFBD process further and investigate what is involved in the fuel-transfer function (3.1.6), and so on.

Note that the numbering of blocks is hierarchical in nature and allows component-level functions to be related through subsystem functions to system-level functions. For example, the numbering system makes it easy to associate function 3.1.3 (Provide fire safety precautions) with function 3.0 (Turn around in 30 minutes).

This example has demonstrated how SRD and functional baseline requirements could be investigated further during preliminary design using the FFBD process. The requirement will be satisfied by whatever subsystem of the aircraft design is responsible for fuel-related function and performance. This may be called the Fuel Subsystem. Determining what physical subsystem will satisfy each requirement is the start of the translation of functional design into physical design, which is one of the key activities of preliminary design.

Some additional points to note:

- *If a refueling time of 17 minutes is impossible to achieve, some additional time may be gained by investigating the other functions*

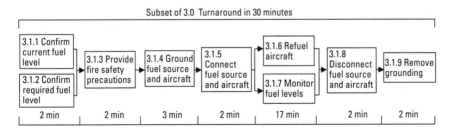

Figure 3.4 An FFBD describing aircraft refueling.

involved in the 3.1 Conduct refueling function. It may be possible to conduct some functions in parallel or it may be possible to speed up some functions. Either way, additional time may be provided to the 3.1.6 function. If it is still impossible to perform the task in the allotted time, at least our traceability allows us to trace back up to the SRD requirement to turn the aircraft around in 30 minutes and revisit this imperative.

- *The FFBD process has also highlighted some interfaces that will need to be considered later in the process. It is worth looking at each of the functions in 3.1 Conduct refueling, and think about the interfaces. Remember that interfaces include human-machine interfaces (HMI), not just physical interfaces. A HMI will exist, for example, to allow the current fuel level in the aircraft to be determined (3.1.1). A physical interface will exist to allow the aircraft to be correctly grounded (3.1.4) and allow the fuel source to connect to the aircraft (3.1.5).*

It is vitally important that all subsystem requirements analysis is performed from the same functional baseline or system specification. Although this sounds like an obvious requirement, different functional baselines have been known to exist on projects, resulting in incompatible and conflicting preliminary and detailed designs. This reinforces the need to review the system specification and the TPMs before embarking on preliminary design.

Once satisfied with the level of decomposition, each functional block can be considered independently, as was the case during system-level requirements analysis. This consideration should focus on the necessary inputs to the function, the expected outputs from it, and any parametric requirements associated with the function. Some thought may also be given to how the function may be performed in reality and what constraints exist in terms of the function's design.

3.3   Requirements Allocation

As illustrated in Figure 3.5, requirements allocation refers to the process of grouping or combining similar functions and parametric requirements (identified during the requirements analysis) into logical subdivisions. We have done this before during conceptual design, where the groups were based on the RBS headings. During preliminary design, the groups are based around a

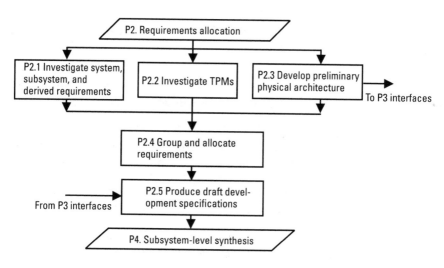

Figure 3.5 The requirements-allocation process.

preliminary physical architecture formulated by the designers, rather than functional RBS headings. The groups of similar functions then help the designer to determine the design of the major subsystems and ultimately the components that are required to make up the system. This represents the translation from functional design to physical design. These subsystems and components are referred to as *configuration items* (CIs) and are selected and designed to perform the group of functions that are assigned to them. It is the assignment or allocation of functions to physical CI that is the major focus of requirements allocation.

The first step in requirements allocation during preliminary design is to investigate each of the subsystem-level requirements and the associated requirements and TPMs flowing from subsystem analysis and group them into logical sets or groups based on a preliminary physical architecture. The logic used relies on experience and expertise in the relevant design domain. In our example, this domain is the aerospace design domain. That is, to perform requirements allocation during preliminary design in our example, we rely on knowledge of the likely physical composition (subsystems) of the aircraft system. Even though the detailed design is yet to be determined, experienced designers have the elements of a preliminary design in mind.

Let's now look at the subsystems that comprise any aircraft (remember that our aircraft example has been pitched at a sufficiently general level to allow anyone familiar with aircraft to participate). First, it is worth looking at what the aircraft industry calls the airframe, which consists of a range of subsystems:

- The undercarriage subsystem includes the structures, tires, and brakes necessary to support the aircraft while on the ground.

- The wings/fuselage subsystem includes all the components necessary to accommodate the remainder of the subsystems and to form the aerodynamic portion of the aircraft system.

- The fuel subsystem includes all the tanks, refueling/defueling receptacles, fuel pumps, and fuel lines necessary to support the engines.

- The hydraulic subsystem includes all the hydraulic pumps, lines, and accessories needed to provide hydraulic power to all other aircraft systems, such as flight controls and undercarriage.

- The flight control subsystem includes all the primary and secondary flight controls necessary for the control of the aircraft system in the air and on the ground.

The next major subsystem of any aircraft system is the engine, consisting of the engine itself and all of the components that come together to form each individual engine, such as turbines and gearboxes. The engine will also require an engine management subsystem, which would include all of the peripheral controls necessary to manage the engine during startup, operation, and shutdown.

All aircraft need systems to assist in the control of the aircraft and perform functions such as electrical generation and distribution, communications, and navigation. Collectively, these functions are normally performed by avionics (which is a term meaning aviation electronics). Most modern avionics subsystems comprise the following:

- A computer (or collection of computers) that includes the hardware and software necessary to perform all aircraft operational functions;

- Navigation subsystems that include the hardware and software necessary to perform the navigation functions for the aircraft;

- Communications subsystems that include the hardware and software necessary to perform the communications functions for the aircraft;

- Automatic flight-control subsystems that include the hardware and software necessary to perform the flight-control functions, such as automatic pilot;

- An electrical generation and distribution subsystem responsible for the generation and distribution of all necessary electrical power to the aircraft systems.

The aircraft system in our example is to be used to transport passengers and freight, so it also has some particular requirements relating to the interior of the aircraft. The interior will comprise the following:

- The cockpit, which includes all of the structures and instrumentation to support the air crew in their operation of the aircraft;

- The passenger space, which includes all of the structures and subsystems necessary to support the passengers throughout their flights;

- The cargo space, which includes all of the structures and subsystems necessary to support the cargo throughout its flights;

- The galley space, which includes all of the structures and subsystems necessary to support the provision of food, drink, and refreshments to crew and passengers throughout their flights.

Each of these major subsystems needs to be considered individually during preliminary design. These subsystems are called CIs because they represent hardware, software, or a combination of both designed to satisfy an allocated group of functions and requirements. As the name suggests, the configuration of each CI is managed as a separate item.

The selection of physical design items and their designation as CIs is a major part of the overall configuration management process known as *configuration identification* (dealt with in more detail in Chapter 5). In general, though, items may be identified as CIs because of the following:

- The complexity and criticality of design;

- The risk associated with the design;

- Costs associated with design, development, or procurement;

- Commonality with other designs or subsystems.

Cis can vary in size and complexity from an aircraft, ship, or electronic system to a test meter or round of ammunition [1].

So now we know the functional requirements for the aircraft system as they are grouped functionally in the populated RBS. We also have decided what the aircraft system is to consist of physically in terms of CIs. What is left to do is to connect the dots and relate the functional requirements to the physical aircraft CI architecture. This is the process of requirements allocation as performed in preliminary design and represents the translation from a functional to a physical architecture. Requirements allocation tells us exactly what each CI needs to do in terms of function and performance.

A proposed allocation is shown in Figure 3.6. On the left-hand side, we see the requirements of the aircraft system (functional representation) and on the right-hand side we list the major subsystems, or CIs (physical representation). The two sides are related by the multitude of lines in the middle.

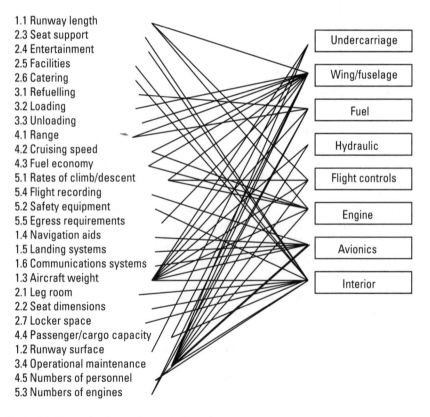

Figure 3.6 Example of grouping and allocation.

Figure 3.6 illustrates one method of showing functional to physical relationships, which becomes unreadable very quickly. A more useful way of showing the same information is via a matrix sometimes called an *allocation matrix* or *traceability matrix*. The information in Figure 3.6 is shown in Figure 3.7 in the form of an allocation matrix.

The RBS functional requirements (representing the functional design of the aircraft system) are shown on the left-hand side of the matrix. The physical design is shown across the top of the matrix as a series of subsystems or CIs described previously. The designers of the individual CIs can clearly identify what requirements and functions have been assigned to their CIs. The designers derive a host of additional subsystem functions and requirements for their individual CIs to ensure that the CIs will be capable of meeting the assigned system- and subsystem-level requirements. These additional requirements are known as derived requirements, as they have been derived from the need to satisfy some higher level requirement (an example of requirements flow down). The higher level requirements and the derived requirements for each CI will form the basis of the technical specification for that CI. The exact form of the CI technical specification depends on the design alternative selected to produce the CI, but for developmental CIs, the specification is usually called a *development specification*.

The allocation matrix concept is also a very powerful tool when changes to either the functional or physical design are proposed. The impact of making changes to particular functions or CIs can be determined very quickly.

From the allocation matrix, it is clear to see that some functional requirements are assigned to more than one physical CI. In these cases, the collection of CIs is jointly responsible for satisfying that requirement. It is worth investigating some examples to explain why this is often the case.

Example 3.2: Engine CI

Weight is a good example of a system-level requirement that has an impact on the design of all of the CIs. In the aircraft example, we are limited in weight due to the requirement to operate from Class X airports (SRD 1.0). Because each CI contributes to the entire weight of the aircraft, each CI is allocated a portion of that entire weight. For example, the Engine CI may be allocated 15% of the maximum entire aircraft weight, and design of the engines must be achieved within that allocation to ensure the entire aircraft weight is acceptable. To understand what else they must achieve with their

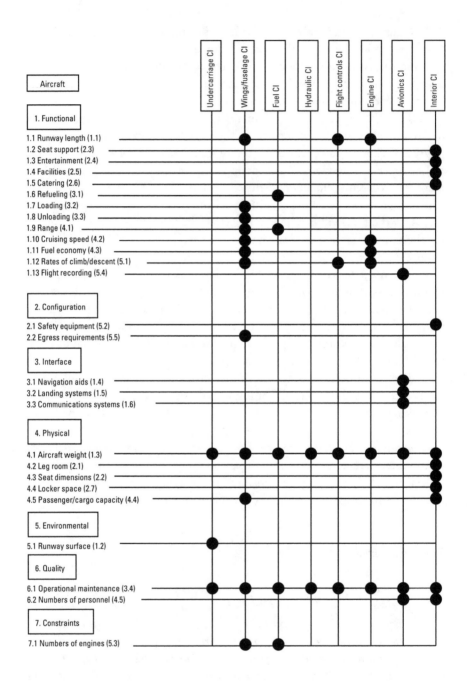

Figure 3.7 Example of an allocation matrix, showing allocation of RBS functional requirements to physical CIs.

engine design, the designers refer (vertically) to the allocation matrix and identify that their engine design must be capable of the following:

- *Propelling the aircraft sufficiently to takeoff on a runway of a certain length;*
- *Achieving a certain fuel efficiency;*
- *Achieving and maintaining a minimum cruising speed;*
- *Enabling the aircraft to achieve minimum rates of climb and descent;*
- *Being maintainable and reliable enough to allow operational-level maintenance to be performed in a very short amount of time.*

From the requirements that have been allocated to the engine CI, the designers derive requirements specific to the engine. For the engine CI, designers may arrive at the type of engine to be used (for example, a jet turbine versus some other type), the required levels of thrust and other performance imperatives, air intake requirements, electrical requirements, mounting and interface requirements, and so on. These derived requirements and allocated system-level requirements form part of the development specification for the engine CI that, when complete, is known as the allocated baseline for that CI.

The process of allocation has ensured that each functional requirement (shown on the left of the allocation matrix in Figure 3.7) has been assigned to at least one CI for satisfaction. By working horizontally in the matrix, we are effectively working from functional to physical design (top down). By working vertically in the matrix, we can appreciate the responsibility each CI has for achieving allocated requirements (bottom up). Bottom-up analysis shows how the functional requirements of the system are driving the design of each of the CIs (not the other way around).

It is now appropriate to ensure that the TPMs determined during conceptual design have been allocated to the various subsystems in the system design. During this stage of the requirements allocation effort, the system-level TPMs are allocated to each of the CIs. The TPMs help to determine specific qualitative and quantitative parameters that must be met by the CIs so that the system-level TPM is met. These parameters are sometimes called *design-dependent parameters* (DDPs). The TPMs and DDPs assigned to the CIs are included in the specifications for the units, bearing in mind that these units will either be procured (off the shelf), designed from scratch, or a combination thereof. Either way, the units must meet the TPMs and DDPs determined during conceptual design and assigned during preliminary design.

It is important to note the advantages of the top-down design approach as borne out by the selection of equipment to meet the derived requirements, TPMs, and DDPs assigned during preliminary design. The subsystems procured or designed for the system are done so with the system-level functional requirements, TPMs, and DDPs in mind. These subsystems are also procured or designed with a good understanding of the major interfacing and integrating issues. By following the top-down approach, the subsystems stand a very good chance of being compatible with one another and combining to meet successfully the functional requirements of the system.

3.4   RBS Versus WBS

Now that the idea of grouping and allocating functional design requirements to physical design CIs has been explained, it is worth revisiting the relationship of the well-known WBS to the systems engineering process. The previous examples have shown the process of establishing the functional design of the aircraft system and the eventual translation of that functional design to physical design. The functional design was articulated in the RBS, and the physical design was documented graphically using an allocation matrix. The concept of a WBS is not far removed from the idea of the allocation matrix. The WBS documents the work packages and products necessary to produce the aircraft system. The work packages and products necessarily include all the CIs listed in the allocation matrix example. In addition, the WBS adds to the CIs the work required to design, develop, integrate, and test these CIs. The breakdown of the work and products in the WBS is based largely on project management imperatives and concepts such as cost and schedule management and earned value systems (if needed). To that end, there is a relationship between the WBS and the RBS that is not dissimilar to the relationship between the CI list and the RBS shown in Figure 3.7.

An example of an Aircraft System WBS [2] is shown in Figure 3.8, including an illustration of the relative location of the CIs (from the allocation matrix). Note that although the CIs do make up a significant portion of the WBS, there is substantial amount of additional work documented in the WBS. Note also that the WBS is similar in structure to the RBS, although the WBS is organized physically, not functionally.

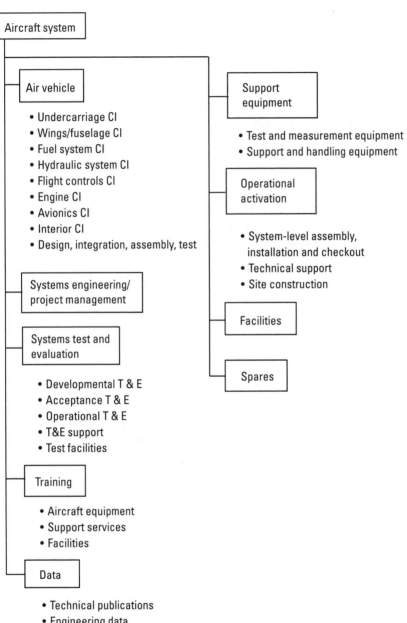

Figure 3.8 Aircraft system WBS [2].

3.5 Interface Identification and Design

During the selection of the CIs comprising the system, the interfaces between the CIs are identified, as illustrated in Figure 3.9.

Identification of interfaces is a critical part of preliminary design as interfaces not only determine successful operation of the system once integrated, they also place additional limitations and requirements on the design of the individual CIs.

The task of integrating different system elements is managed primarily through a document called the *interface-control document* (ICD), which contains sufficient details to completely define the interfaces between the different subsystems. A primary source of information for identifying interfaces between CIs is the functional flow block diagrams produced as part of the requirements analysis effort in the conceptual and preliminary design phases.

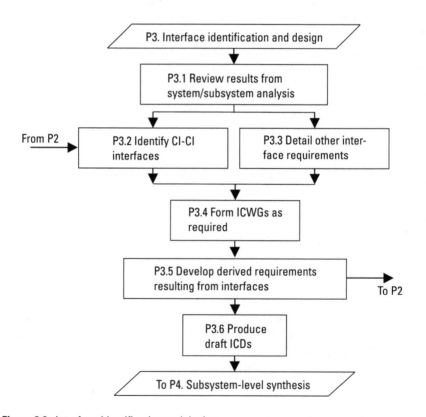

Figure 3.9 Interface identification and design.

The ICD contains various types of information, depending on the nature of the interface. The following paragraphs provide examples of some of the different types of interfaces that exist. Note that to completely define some interfaces, an ICD may require information on more than one type of interface.

- *Physical interfaces.* Physical interfaces are normally the most obvious of interfaces because they need to exist in real terms. They may be pipes through which a fluid or gas flows or they may be a fiber-optic cable passing digital information between two computer systems. In defining the physical interface, factors such as mechanical connection type and specification, physical size of the interface (such as diameter of pipe), and layout of the interface are defined. Physical interfaces normally coexist with one of the other types of interfaces.

- *Electronic interfaces.* An electronic interface is the name given to the flow of electronic signals (analog or digital) between two points. These signals may flow over a physical interface such as a cable or fiber, in which case the physical cable or fiber interface is also defined. Alternatively, the interface may make use of radio frequency transmission such as microwave or satellite to pass the information. The characteristics of the data flow between the two points needs to be completely defined. The information requirements for this definition depend entirely on the nature of the interface but may include factors such as modulation method, frequency, data format, and data rates.

- *Electrical interfaces.* Electrical interfaces are normally associated with a physical interface of some sort such as a power cable or data cable. The type of information required in this sort of interface will include voltage levels, voltage type (for example 110V 60Hz), and any associated factors such as fault tolerance requirements. In the case of a data cable, pin assignments and descriptions may be required. If there are any electromagnetic compatibility issues relating to the interface such as shielding and grounding, these requirements will also be defined.

- *Hydraulic/pneumatic interfaces.* Hydraulic and pneumatic interfaces are the mechanical equivalent of the electronic or electrical interfaces already described. A physical interface definition will coexist with this type of interface. Additional information such as flow rate, pressure, temperature, and fluid type will be required to completely define the hydraulic and pneumatic interfaces.

- *Software interfaces.* Software interfaces refer to passing of data between different computer software CIs. This may occur within the one piece of hardware using internal hardware busses or it may be between two different pieces of hardware. If the latter is the case, a physical interface definition is also required to define the physical link between the two pieces of hardware. Software interfaces require information such as expected inputs, generated outputs, format of the messages, and any protocols used during the transmission. Protocol definition includes data rate, error detection and correction, and start and stop codes.

- *Environmental interfaces.* Environmental interfaces are sometimes not considered true interfaces, but they form the interface between the system and its external environment. In some case, the environment provides the system with important information, such as air data in the case of an aircraft. The environment may also provide the system with challenging interfaces such as structural (vibration, shock, acoustics), thermal (ranges of temperatures), magnetic (flux densities), and radiation (flux densities and types). All environmental interfaces need to be considered and documented to ensure that the most important integration of them all (system with its operational environment) is successful.

Example 3.3: Interfaces and CI Design

Example 3.2 looked at the design of the Engine CI using the allocated set of functional requirements to drive the derivation of additional requirements. The allocated requirements are not the only set of requirements that will impact on the design of the engine, however.

The engine must interface with a number of other CIs, including the Fuel CI. An interface between the Fuel and Engine CIs will be identified and documented during preliminary design (see also integration management in Section 5.7). The documentation of this interface includes details of the physical interface between the two CIs and also describes critical performance factors such as maximum flow rates, pressures, and physical means of connection. Both the Engine CI designers and the Fuel System CI designers have specific requirements from the interface. To that end, an interface-control working group (ICWG) consisting of representatives from both CIs is usually formed to participate in the design of the interface.

It is clear that characteristics of the interface design such as flow rates, fuel pressure, and physical means of connection form additional constraints on the design of both the Fuel and Engine CIs. The constraints produce a number of derived requirements that must be included in the relevant specifications for the CIs.

3.6 Subsystem-Level Synthesis and Evaluation

Once the allocation of requirements has been performed, it is time to continue the familiar loop of analysis, synthesis, and evaluation and concentrate on synthesis. Subsystem-level synthesis results in a preliminary design consisting of hardware, software, personnel, facilities, data, and so on to meet the subsystem-level requirements. Figure 3.10 illustrates the process.

3.6.1 Review Sources of Subsystem Requirements (Figure 3.10, process P4.1)

The allocation process that has grouped similar functions together into logical groupings has already completed the initial analysis. Designers responsible for

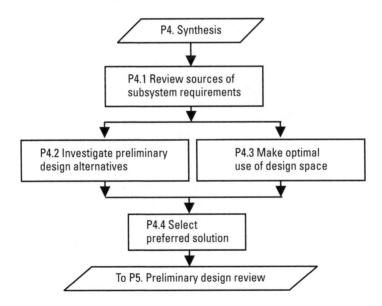

Figure 3.10 Subsystem-level synthesis.

each CI have continued the analysis to an additional level of detail necessary to support the subsystem-level synthesis. Technical specifications and ICDs have been produced to detail the functions, performance, and interface requirements for each CI. The initial synthesis is conducted to determine a preliminary design that satisfies the functional groupings using one of the three broad design options available. As a first step in the synthesis process, the specification and ICDs for each CI should be reviewed, as it is these specifications that must be satisfied by subsystem synthesis.

3.6.2   Investigate Preliminary Design Alternatives (Figure 3.10, process P4.2)

There are three broad design alternatives available including COTS, modified COTS, and developmental items.

- *COTS.* COTS equipment, as the name suggests, refers to pieces of equipment that are commercially available. Military off-the-shelf equipment is another category of off-the-shelf equipment that has been included in the category of COTS for the purposes of this discussion. If the COTS equipment meets or exceeds the requirements detailed in the development specification, there may be advantages to using COTS equipment. The most obvious advantages are that the equipment is likely to be readily available (no delay), it is also almost certainly cheaper than the alternative, and maintenance and support will probably already be in place for the item. The disadvantages often associated with COTS items include problems with form, fit, or function, including inappropriate size, shape, weight, or color, or excessive or insufficient functionality. COTS items will ideally be described by a product specification (type C) delivered with the equipment but in all likelihood the documentation set available will be substantially less detailed than a product specification. The documentation does, however, need to be detailed enough to support tasks such as integration, maintenance, and support. The need to qualify new pieces of COTS equipment in the system's operational environment may also weigh in on the decision regarding the use of COTS. Qualification is a time-consuming and expensive exercise and may reduce considerably the benefits associated with the use of COTS equipment. For example, before using a COTS radio system in our aircraft example, the radio system would need to be proven or qualified using a set of standard test procedures to ensure the radio could survive and function under the temperature, vibration, shock, and humidity extremes associated with aircraft operations.

- *Modified COTS.* COTS equipment that meets a majority of requirements contained in the development specification for a CI may be modified to better suit the needs of the designer. Similar advantages will exist with modified COTS as for COTS, however some maintenance and support may be voided if the item is modified. Another point to consider is the effort involved in modifying the item. This effort is easily underestimated and the modification process can take a lot longer and cost a lot more money than initially estimated. Modified COTS equipment may be described in a number of ways. One way is to describe the original COTS equipment using the standard COTS product specification (type C) and describe the necessary modifications to the COTS item with a COTS modification specification. A COTS modification specification may include a combination of process specifications (type D) to describe how to modify the COTS item and materials specifications (type E) detailing the materials needed to carry out the modification. These specifications are described in more detail in Chapter 5.

- *Developmental items.* If suitable COTS equipment is not available, the designer may opt to design and develop the item from the ground up to meet the specific requirements and characteristics detailed in the relevant development specifications. An item developed from the ground up is likely to match the desired criteria precisely in terms of form, fit, and function, however the effort involved in this development will not be insignificant. Maintenance and support issues will also need to be considered and resolved when items have been designed specifically for the system. Hardware and software development specifications (type B) and hardware and software product specifications (type C) will describe developmental items in progressively more detail. These documents will contain information on the specific item and its particular interfaces. Hardware items may then require assembly process specifications (type D) and subassembly material specifications (type E) to complete the description.

Other factors that impact on the designer's decision to use COTS, modified COTS, or specially developed items include the specific functions to be performed by the item, the availability and stability of the current technology, size of the market, supportability, and cost. The process of grouping functions together and then allocating those functions to subsystems is shown in Figures 3.6 and 3.7. Note how the allocation process has effectively specified the functional performance of the subsystems.

3.6.3 Make Optimal Use of Design Space (Figure 3.10, process P4.3)

During the selection of the different subsystems making up the system, it is important to remember that it is the system performance that is of vital importance, not the individual performance of the subsystems and components. The reason system performance needs to remain the focus of the preliminary design is that to achieve optimal system performance, the performance of the subsystems may need to be suboptimal. A simple example illustrates this concept.

Example 3.4: System/Subsystem Optimality

A traffic management system (Figure 3.11) consists of a set of road subsystems (AA, BB, CC, 11, 22, 33). For each road subsystem to operate optimally with maximum traffic flow, all traffic lights on that subsystem must remain green. While that may be a reasonable design decision in the context of any single subsystem, this situation will clearly result in suboptimal performance at the system level. Note that this example has been simplified to illustrate the concept of system/subsystem optimality. Most design problems are multidimensional, representing nonlinear problems that require multidisciplinary teams to solve.

To gain optimal system performance, trade-offs need to be considered during the selection of the preferred design approach. These trade-offs take

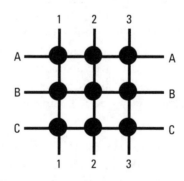

Figure 3.11 A road system comprising a number of subsystems.

into account the performance requirements of the different subsystems and that of the system, and also consider any design space available for trade-off purposes. This helps ensure that the most important system-level perform-ance measures (as ranked during conceptual design) are met first followed by the less important parameters.

Example 3.5: Aircraft System Design Space

The aircraft system development has reached a stage where the design of the CIs can commence. This example investigates the design of the Interior CI and, in particular, the actual passenger seat. The example demonstrates the concept of design space and its use during preliminary design (and subse-quent design activities).

2.2 Minimum Seat Dimensions. We know that the seat design is of critical importance to the success of the aircraft system. By using traceabil-ity, we know that the seat dimensions are a component of the overall SRD aim to provide "class-leading levels of comfort." Let's say for this example that the minimum acceptable seat size is 576 in^2. Seat dimensions have been allocated to the Interior CI. When speaking to the designers responsi-ble for the seat design, they explain that the seat dimensions are critically limited by the number of passengers on board. Naturally, the more passen-gers, the smaller each seat becomes. We have a minimum requirement for the number of passengers that the aircraft system must be capable of trans-porting (4.4 Passenger/Cargo capacity). For the purposes of this example, the minimum acceptable passenger load is 200 people.

The relationship between seat dimensions and passenger load (as de-termined by the designers of the Interior CI) is shown simplistically in Figure 3.12.

Based on Figure 3.12 alone, we feel very confident that we can meet both requirements. In fact, we can either push the number of passengers up to 300 (exceeding requirement 4.4) and still meet the seat dimension re-quirement, or we could meet the passenger requirement and provide each passenger with a much bigger seat. In short, we have some design space. The commercial viability of the aircraft may be deemed more important than exceeding passenger comfort levels. If this is the case, then the decision may be to maximize the number of passengers on the aircraft and opt for 300 passengers with 576-in^2 seats.

Provided we keep within the design space available, we will meet or exceed all requirements. The aim in using design space is not to exceed some requirements at the expense of others, but rather to strike a balance and use

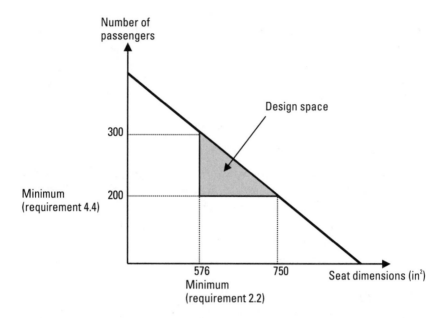

Figure 3.12 The design space represented by number of passengers and seat dimensions.

the design space to at least satisfy all requirements. To use design space appropriately, however, the stakeholders must be consulted. Designers do not necessarily own the design space; the stakeholders do. In this example, the commercial stakeholders were given priority over the stakeholders vying for additional seat size.

1.3 Maximum Aircraft Weight. *To meet the SRD requirement to operate from all Class X airfields in the world, a maximum aircraft weight was applied. For this example, we assume that the aircraft weight is not to exceed 100,000 lbs. Weight was allocated to all CIs, including the Interior CI. Figure 3.13 illustrates how passenger numbers impact on the entire aircraft weight.*

The design space relating to passenger numbers has now shrunk somewhat from that described in Figure 3.12. Based only on seat dimensions, the aircraft was capable of carrying 300 passengers. However, when entire weight is considered, only 250 passengers can be carried. The designers of the Interior CI will be happy because they will now be able to exceed the minimum seat dimensions set for them (see Figure 3.12). Note also that we have a minimum achievable entire weight of 90,000 lbs if we only carry the minimum acceptable number of passengers.

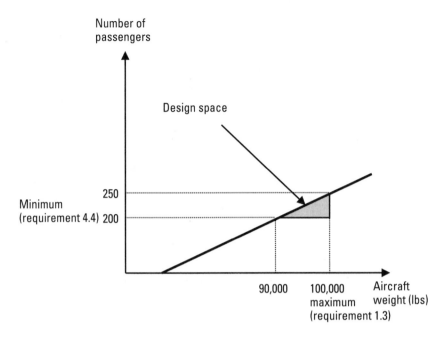

Figure 3.13 The design space represented by number of passengers and aircraft weight.

Often this type of analysis reveals unavoidable conflicts between functional and derived requirements and the design of individual CIs. For example, the designers of the Engine CI may identify a COTS engine that meets or exceeds all allocated and derived requirements except for the minimum thrust requirement. The engine does not produce sufficient levels of thrust to support the allocated requirements of 1.1 Runway length, 4.2 Cruising speed, and 5.1 Rates of climb/decent for a 100,000-lb aircraft. The designers are attracted to the COTS engine because it offers all of the classic advantages of COTS design; reduced schedule, cost and technical risk and well-established logistics support. Further to this information, the designers indicate that it is not technically feasible to provide the necessary levels of thrust from an aircraft engine using current technology for a 100,000-lb aircraft. The engine designers illustrate their case with Figure 3.14, which shows thrust versus weight necessary to achieve the allocated requirements of runway length, cruising speed, and rates of climb/descent.

Figure 3.14 clearly shows that the engine will be incapable of supporting the performance requirements for an aircraft that exceeds 85,000 lbs. In isolation, this is not a problem because 100,000 lbs is, after all, a

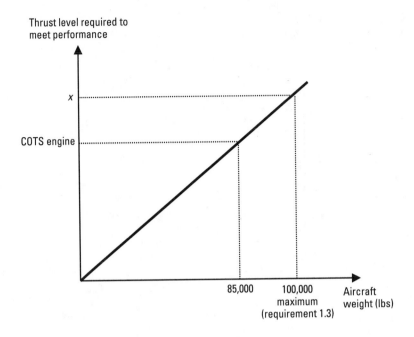

Figure 3.14   Revised power-versus-capacity function.

*maximum and an 85,000-lb aircraft will obviously meet that require-
ment. However, when passenger load is revisited, it is seen from Figure
3.14 that the 85,000-lb upper limit is not sufficient to meet the minimum
passenger load of 200. This is an important conflict in the requirements.
Traceability now becomes critical as the systems engineer traces the requirement for minimum passenger load back to the functional baseline (system-
level) requirements from which that requirement was derived. The performance requirements came, in part, from regulatory requirements that
are not negotiable. In this simple example, it is easy to remember that passenger load was related to commercial viability. The systems engineer must
alert the owners (the relevant stakeholders) of this requirement as soon as
the conflict becomes apparent, seeking guidance, clarification, and prioriti-
zation. Based on the feedback, design decisions can be made while keeping
the owners informed. In reality, the system-level designers may be able to
save some weight in other areas of design (by referring to the design space) to
allow the minimum passenger load to be achieved while keeping the entire
aircraft weight within the 85,000-lb limit set by the engine design.*

*This exercise has demonstrated the concept of design space and trade-
offs to meet the original requirements of the system. The example has also*

demonstrated the concept of suboptimality where subsystem design must not take priority over achievement of system-level performance requirements.

Trade-offs within design space must be applied to all of the subsystems making up the system (as defined during the subsystem functional analysis and requirements allocation effort). Each subsystem affects different parts of the functionality and has different TPMs and DDPs assigned to it. To that end, each subsystem has its own set of trade-offs and design spaces within which it must operate.

3.6.4 Select Preferred Solution

The purpose of allocation of requirements is to assign some performance guidelines to each of the subsystems making up the system to allow design engineers to select the subsystem that will be compatible with, and contribute to satisfying, the system-level requirements. This process is iterative, involving the familiar analysis, synthesis, and evaluation activities. It may not be possible to meet all of the allocated requirements exactly on the first run through the design. The design (or synthesis) is evaluated to determine the level of success in meeting the requirements. Shortfalls are identified and the process must be repeated with different variables being traded off until as many of the allocated requirements as possible are met.

Once the preliminary design process is mature, the result is a preliminary architecture made up of various units and configuration items of hardware, software, personnel, and so on designed and organized in such a way to meet a series of requirements that have been assigned to each component of the preliminary design. Each subsystem has an individual description of its intended purpose, and the requirements that it must be capable of meeting to fit into the system as a whole called a *specification*. These individual specifications must be detailed enough for a hardware engineer to design and build a piece of equipment (if that is deemed to be an appropriate approach), or for a software engineer to write a piece of software to perform the functionality, or for a logistics person to source a piece of COTS equipment to satisfy the requirements. Clearly, the level of definition for these individual specifications will vary. In the case of the hardware and software design options, these specifications are called *development specifications,* or *type B specifications.* For the COTS equipment, the specifications can move straight to the *product specification,* or *type C specification,* for the specific piece of COTS to be procured.

This allocation of requirements to subsystems and the development of detailed product and development specifications are called the *allocated baseline* and is the major product of preliminary design.

3.7 Preliminary Design Review

As with all formal design reviews, the PDR is aimed at ensuring the adequacy of the preliminary design effort prior to allowing the design focus to shift onto detailed design. PDR is designed to assess the technical adequacy of the proposed solution in terms of technical risk and the likely satisfaction of the functional baseline. PDR also investigates the identification of subsystem interfaces and the compatibility of each of the CIs. The major PDR activities are illustrated in Figure 3.15.

As with all reviews, PDR must be conducted only when the customer is satisfied that the system design has progressed enough to justify holding the design review. Prior to commencing PDR, the customer must be confident that the system architecture is in place, with all CIs identified and documented to an appropriate level of detail in development specifications.

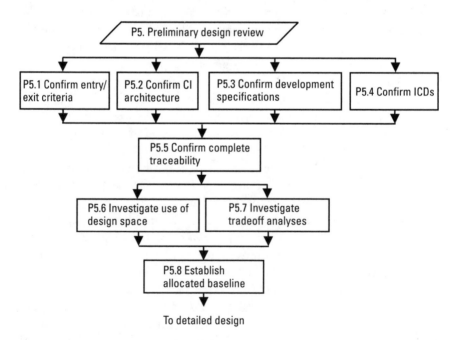

Figure 3.15 Major PDR activities.

Interface identification and design should also be confirmed by investigating the state of the ICDs prior to the review. The customer also needs to be satisfied that other entry criteria such as financial and project management criteria have been satisfied.

The major task during the PDR is to investigate each of the CIs that have been determined during preliminary design. This process is aimed at verifying that all of the overall system-level requirements are being met. In addition, the PDR investigates the results of the subsystem requirements analysis, the requirements allocation process, and the various trade-off studies conducted during the synthesis stage. Again, the results of these activities should be seen to address all requirements. Any deviations from the requirements are noted and corrective action determined and initiated.

At the conclusion of a successful PDR, the participants should be confident that all of the functional baseline requirements are being adequately addressed by the subsystem design. The functional and physical interfaces that need to be in place between the CIs themselves and between the CIs and the external environment have been identified and documented ready for detailed design. Participants should also be satisfied that the allocated baseline is in a fit state to be used during detailed design and development. Any additional exit criteria listed against PDR should also have been adequately demonstrated. The successful conclusion of PDR results in the establishment of the allocated baseline against which detailed design is subsequently authorized. Sometimes an issue is raised during PDR that needs resolution prior to commencing detailed design and development. Either the PDR can be stopped until the issue is resolved or a decision is made to continue with the PDR and the issue is dealt with separately in a different forum.

Two approaches can be taken with respect to PDR. Either a separate PDR can be conducted for each of the major subsystems (or CIs) of the system or a single PDR can be conducted covering multiple CIs. The approach taken should depend on the size and complexity of the system under investigation. The principal documentation reviewed and approved at PDR are the development specifications relating to each of the CIs and the ICDs describing internal and external interfaces. Preliminary product specifications (if available) may also be reviewed at PDR to ensure that the effort is on track.

The aims of the PDR include the following:

- *Evaluation of the preliminary design.* PDR investigates preliminary design activities to ensure that they have been conducted thoroughly.

- *Approval of development specifications and ICDs.* The subsystems defined during preliminary design are documented using development specifications. PDR reviews these documents and approves them prior to detailed design and development.
- *Assessment of traceability.* The review needs to ensure that all of the requirements in the functional baseline have been adequately captured and addressed in the preliminary design. This is the concept of forward traceability introduced in Chapter 1. Backward traceability also needs to be confirmed to ensure preliminary design does not include unnecessary "gold plating."
- *Establishment of the allocated baseline.* The development specifications and top-level design developed during preliminary design form the allocated baseline. This baseline is formally established at the conclusion of PDR and paves the way for the detailed design effort.

Other documents supporting the top-level design are also reviewed during PDR if available. Examples of additional documentation include test plans, preliminary engineering drawings, and production plans.

Particular attention should be paid to high-risk areas, interface requirements, and system-level trade-off analyses. If appropriate, TPMs identified earlier in the system design effort should be discussed and progress traced during PDR.

Endnotes

[1]   MIL-STD-481B, *Military Standard Configuration Control—Engineering Changes (Short Form), Deviations and Waivers*, U.S. Department of Defense, Washington, D.C., July 15, 1988.

[2]   This example is derived from that given in MIL-HDBK-881, *Department of Defense Handbook—Work Breakdown Structure*, Washington, D.C.: U.S. Department of Defense, 1998, which provides an excellent reference to WBS.

4

Detailed Design and Development

4.1 Introduction

The detailed design and development activity continues the development effort and makes use of the functional and allocated baselines developed during conceptual design and preliminary design. The detailed design effort takes these definitions of the overall system (functional baseline) and of the major subsystems (allocated baseline) and starts to consider the design of specific subsystem components. The realization and documentation of individual components is referred to as the *product baseline.*

A number of tasks are undertaken during detailed design and development. The major technical activities include the following:

- Describing the lower-level components making up the subsystems, including software products and personnel requirements (and their interrelationships);

- Defining the characteristics of these items through specifications and design data;

- Finalizing the design of all interfaces necessary to support system integration;

- Procuring the above items off-the-shelf or designing them if they are unique to the system under development;

- Developing a prototype or engineering model of the final system by integrating these items (for the purpose of system-level test and evaluation);

- Redesigning and retesting (as required) aspects of the system prototype that failed to meet the minimum requirements;

- Conducting a critical design review (CDR) to ensure that the design is ready for construction and production.

4.2   Detailed Design Requirements

Specific detailed design requirements must be derived from the system specification (developed during conceptual design) and the various development and product specifications (derived during preliminary design). The content and intent of the specifications developed in detailed design and development is detailed further in Chapter 5. The detailed design specifications contain allocated requirements, appropriate DDPs, TPMs, and associated characteristics that must be incorporated into the design of specific components. The specifications also necessarily contain descriptions of the requirements (functional, physical, and performance) for the interfaces relevant to the subsystems and components. In addition, there may be a requirement to specify the fabrication processes and materials necessary to produce the detailed design.

The specification process is an excellent example of top-down design process, which is responsible for establishing specific requirements at each level in the system's hierarchical design structure. The process evolves through the now familiar iterations of analysis, synthesis, and evaluation until the definition of all system components (documented in the appropriate specifications) is complete. The collection of specifications defining the system's components is called the *product baseline*.

4.3   Designing and Integrating System Elements

Through the process of system and subsystem functional analysis and allocation, all elements of the system and their specific characteristics have been identified and documented. The entire system has now been defined in terms of its whats. The design engineer now needs to use trade-off analyses and other design tools to determine the best way to deliver these whats. That is, the hows need determination if they have not already been determined in an earlier phase.

4.3.1 Detailed Design Process

A typical detailed design process is very iterative in nature and is dominated by reviews and feedback at each stage. The detailed design process is initiated by the completion of the previous phases in the acquisition process (conceptual and preliminary design). The system specification and set of development specifications are provided by preliminary design and enable designers to determine the detailed design and definition of the subsystems and components (via the familiar analysis, synthesis, and evaluation process). These definitions are then reviewed to ensure that the definitions are complete and meet the overall system-level requirements. If the review is negative, the process is repeated. The detailed design process is shown in Figure 4.1.

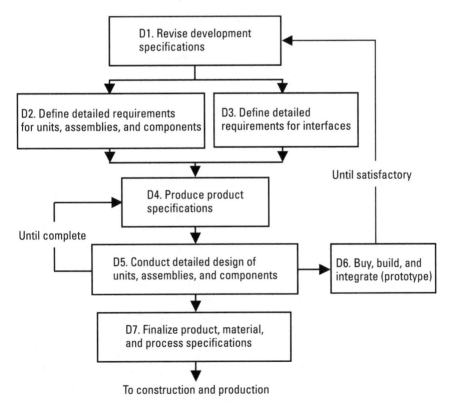

Figure 4.1 Detailed design-and-development process.

The subsystem definitions are then further investigated and broken down into lower level portions of the design (including components, units, assemblies, software, subassemblies, and parts). Again the review process ensures that this process is complete and accurate before the design effort focuses on producing the specifications detailed earlier to document the design effort. The documented system is then procured, modified, or developed and integrated, as discussed in the next section, to produce an initial prototype upon which system-level testing can take place. If the result of this testing is positive, the system can move into construction and production. If not, some recommendations as to system modification will be necessary. Depending on the extent of the modifications needed, the process returns to one of the previous detailed design stages.

4.3.2   Integration

Successful integration relies on accurate and complete interface definition during the preceding phases of the design. The ICD and individual interface requirement documents contained in development and product specifications all play an important role in the integration task. In addition to documentation, cooperation between the various design groups (such as hardware and software) is vitally important. Note that some system components may be completely designed and developed in geographically remote locations and only brought together during the integration stages.

An effective interface definition and design tool is known as the N2 diagram [1]. Although originally a software engineering tool, the diagrams are extremely useful in all types of interfaces. In simple terms, the diagram represents a matrix with system functions running down the diagonal. An example N2 diagram is illustrated in Figure 4.2.

The interfaces are shown on either side of the diagonal. The interfaces are identified in both directions because an interface may represent an input to one function and an output from another. If no interfaces exist between certain functions, the relevant box in the diagram is left blank.

The N2 diagram can also use physical items in place of functions. During preliminary design, these physical items will be the CIs that have been identified and the interfaces will represent interfaces between the CIs. As the design becomes more detailed during detailed design and development, these interfaces will be refined further. In this way, the N2 diagram can support the top-down identification and design of interfaces throughout the development life cycle from very high-level functional interface identification during

Function 1 (F1)	F1 → F2	F1 → F3	F1 → F4	F1 → F5
F2 → F1	Function 2 (F2)	F2 → F3	F2 → F4	F2 → F5
F3 → F1	F3 → F2	Function 3 (F3)	F3 → F4	F3 → F5
F4 → F1	F4 → F2	F4 → F3	Function 4 (F4)	F4 → F5
F5 → F1	F5 → F2	F5 → F3	F5 → F4	Function 5 (F5)

Figure 4.2 A sample N2 diagram.

conceptual design to very specific and detailed interface design during detailed design and development.

An example of a partial CI-level N2 diagram for the aircraft system is shown in Figure 4.3. Note the inclusion of External systems to allow the identification of interfaces with external systems and the environment.

Each of the interfaces identified in the N2 diagram shown earlier is considered separately using the ICDs described in Chapter 3. An ICD is established for each interface during preliminary design to document the physical and function requirements for the interface. Detailed design of each interface is conducted during detailed design and development and documented in the relevant product specification. For example, the interface to provide hydraulic pressure to the fuel pumps in the fuel system would be documented in the interface design section of the hydraulic system's product specification. The fuel system documentation would also note that it is expecting hydraulic pressure to be provided by the hydraulic system.

Avionics	• Throttle controls • Electrical power	Fuel pump control		HMI
Mechanical power	Engine		Mechanical power	
Inputs to fuel gauges	Fuel pressure/ supply	Fuel		Current quantity indication
Inputs to hydraulic gauges		Pressure to fuel pumps	Hydraulic	Current quantity indication
Landing and navigation inputs		Refueling equipment	Refilling equipment	External systems

Figure 4.3 Example of a partial N2 diagram for our example aircraft system.

Detailed design has now reached a stage where each CI has been designed completely, as have the interfaces between the CIs and between the CIs and external systems. The individual items can now be procured (in the case of COTS items), procured and modified (in the case of modified-COTS items), or constructed from the ground up (in the case of developmental items). The items can then be integrated in accordance with the overriding system architecture to form the prototype of the final system.

Acquisition may involve procurement of COTS items, modification of COTS items if required, or the construction of development items as detailed in Chapter 3. Integration aims to combine lower level components into progressively higher level subsystems until the system prototype is complete. While the design process has been conducted top-down, the integration process is conducted bottom-up. At each stage of the integration, some form of integration testing is conducted to verify the successful integration against the appropriate level of documentation (that is, the product and allocated baseline). Eventually, when systems integration is complete, testing can be conducted at the system level against the functional baseline. Testing and evaluation play a role in all phases of the systems engineering effort and are described in greater detail in Chapter 5. The integration effort is summarized in Figure 4.4.

4.3.3 Some Detailed Design Aids

Detailed design is a challenging task and fortunately, as technology advances, sophisticated design aids are emerging to help the system designer. These aids allow the designer to realize rapidly and inexpensively different design alternatives for the purpose of comparison and selection. The different alternatives can be traded off against one another to ensure that the most suitable design alternative is selected. This section provides some general examples of design aids used for hardware and software engineering. There is not sufficient space for the inclusion of more specific information.

Hardware engineering has benefited over the last ten years from the explosion in computing power available to the end user. Computer-aided engineering (CAE) and computer-aided design (CAD) are two examples of hardware design tools that allow engineers to perform analysis, synthesis, and evaluation at each stage of the design process without having to resort to expensive and time-consuming physical engineering models and mockups. It should be noted that software models do have some limitations and that physical models and mockups are often still sometimes required to allow designers to verify physical layouts, human-machine interfaces, cable routes, access points, and so on. Models and mockups can also assist related disciplines such as logisticians to verify reliability and maintainability estimates and production engineers to confirm manufacturing processes [2]. Modern CAE/CAD packages allow engineers to develop rapidly three-dimensional models of proposed hardware and to experiment with different layouts and

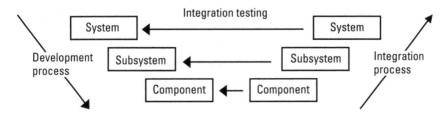

Figure 4.4 Top-down development and bottom-up integration process.

designs. CAE/CAD packages often interface with computer-aided manufacture (CAM) systems to develop parts lists for the final configuration of the design.

In terms of software development, a wide (and ever-increasing) range of development environments exist today to allow software engineers to rapidly mockup user interfaces and functional software for the purposes of design evaluation. Development environments incorporating development tools, debuggers, editors, and compilers are commonplace today. Embedded software (used in microcontrollers) can be developed, debugged, and compiled in a single environment, and the same environment usually incorporates a simulator function to simulate the behavior of the hardware within which the embedded software will operate. Computer-aided software engineering (CASE) tools assist developers in their efforts to design, test, and document software components of systems.

It is not necessary to go into great detail about the design technologies available to modern designers. It is sufficient to understand that technology is helping to make the design process quicker and more likely to arrive at the correct answers.

4.4   System Prototype Development

All of the engineering design effort to this point has been focused on breaking the overall system-level requirements down to a detailed level, assigning these requirements to given unit-level components (hardware, software, facilities, and so on), and then proceeding to designing these individual elements. Analysis has been continually used to ensure that the overall system-level requirements are being met by the design alternatives selected. Analysis forms one of the cornerstones of the systems engineering effort and is an absolutely vital part of the entire process. However, at this stage in the process, there is a need to verify the design concepts and system configuration using the actual system components arranged, combined, and integrated in a "final" state. The arrangement, combination, and integration of the components is sometimes called a system model or prototype. The system-level requirements can now be physically demonstrated using this model, in conjunction with a defined test and evaluation regime. This system-level test and evaluation is vital, considering the concept of subsystem suboptimality introduced in Section 3.6.

Prototypes can vary from very complete and almost final versions of the system in terms of form, fit, and function through to models that only represent the functional characteristics of the final system.

Example 4.1: Aircraft System Prototypes

A prototype of our avionics system may be no more than a series of networked personal computers running a mature version of the avionics software suite, sometimes called an operational flight program (OFP). This network of computers may be operating in an office environment in the center of a major city.

Conversely, the same avionics system development may result in a prototype that consists of the final aircraft hardware mounted in a mockup of the aircraft equipment bay and wired together using the actual wiring looms to be used in the aircraft installation. Actual aircraft instrumentation may be used and simulated parameters such as basic air data may be used to stimulate the model.

The type of prototype used will depend very much on the nature of the project and system. In the case of safety or mission critical systems, the system prototype must offer stakeholders high levels of confidence in the robustness of design. For example, aircraft avionics now impacts on the very airworthiness of aircraft, so it is easy to justify very sophisticated prototypes to verify the correct performance of aircraft avionics software under a range of conditions prior to testing the avionics subsystem in an actual aircraft. On the other hand, a prototype designed to confirm the layout of the aircraft galley is likely to be quite a simple arrangement.

4.5 Detailed Design Reviews

As with each of the preceding design phases, a design review (or a series of reviews) is normally conducted at the conclusion of detailed design and development to ensure that the system design is complete prior to construction and production.

4.5.1 Equipment/Software Design Reviews

A number of informal design reviews are conducted throughout the detailed design phase. These design reviews, sometimes called equipment/software design reviews, are focused on particular items of software and hardware to ensure that the specific design approach meets the requirements. These reviews investigate the drawings, computer programs, material lists, and the necessary specifications to thoroughly review the item under investigation.

Associated documentation, such as trade-off reports, that justify and explain the design approach will also normally be investigated during these reviews. If component-level prototypes or models have been produced and tested, the results of the testing and any functional discrepancy rectifications will also be tabled.

All equipment/software design reviews will be completed prior to the major formal review of the phase, which is the CDR.

4.5.2   Critical Design Review

CDR is the final design review prior to the official acceptance of the design and the subsequent commencement of construction and production activities. The result of the successful completion of CDR is the establishment of the product baseline, which is the effective freezing of all design activity. Only discrepancies in the system design identified during testing result in further design activity following CDR. As with the PDR, CDR needs to be conducted on all of the CIs of the system (hardware and software). Whether a separate CDR is conducted on each CI or a single system-level CDR is conducted (reviewing all CIs at once) depends on the size and complexity of the system under investigation.

CDR aims to demonstrate that the CIs under review are able to satisfy the functional and performance requirements allocated to them in the product specifications. CDR also confirms the compatibility of the CIs with themselves and with other parts of the system including other equipment, facilities, and personnel. At this time, test and evaluation activities are reaching a critical stage, and CDR confirms that strategies, procedures, and support are in place to ensure a comprehensive test and evaluation process. CDR is also the forum in which plans for construction and production of the system are evaluated and approved.

A key aspect of CDR is the investigation of the TPMs that were identified during conceptual design and whose progress has been tracked and monitored ever since. All TPMs should be either exceeding or meeting the necessary levels by CDR. By this mature stage in the design, there should be no surprises as far as TPMs are concerned. Departures from the accepted results should be noted and corrective action attempted, although at this stage in the project any departures are normally difficult to rectify.

As with all design reviews, a clear demonstration of readiness should be provided prior to commencing the review via the satisfaction of all entry criteria. All necessary documentation must be received and reviewed in advance of the design review. Examples of documentation that would be expected to be reviewed prior to CDR includes revised interface documentation such as ICDs, specifications such as development specifications and product specifications, test and evaluation procedures, relevant technical data including assembly diagrams and drawings, installation drawings and schematics, software source code listings, and comments. Reviewers should be convinced that all necessary equipment or software reviews have been conducted prior to CDR. There is little point conducting a system CDR prior to the completion of all CI-level CDRs. The system architecture and detailed design should be mature prior to commencing CDR as CDR marks the effective freezing of design activities. If design effort is continuing, then the design is not ready for the CDR. This judgment will be possible following receipt and review of the documentation set already described. Once satisfied that the system design is ready for CDR, the review can commence.

For CDR to be considered a success, the reviewers need to be satisfied that the functional and performance requirements allocated to each CI will be satisfied by the design for the CI presented at the review. This provides confidence that the requirements contained in the functional baseline will be achieved.

Out of CDR may flow the requirement to update the functional baseline or allocated baseline. The product baseline is established by successful completion of CDR, and the system is then ready to move into construction and production in accordance with approved construction/production planning.

The timing of CDR within detailed design and development is critical. If the review is planned too early in the process, the design may be immature and may result in a number of postreview changes. Post-review changes impact on the production planning and may introduce additional risks in the latter stages of the acquisition phase. Obviously, holding CDR too late in the project adversely impacts on the production schedule and may result in testing and delivery delays.

It is likely that some integration and prototyping will have occurred since the last review (PDR), and CDR investigates and assesses the success of this effort as a key indicator of design maturity.

The aims of the CDR include the following:

- *Evaluation of the detailed design.* A great deal of effort has been put into establishing the detailed design of the system during this phase. This design is documented in various forms including the mature product specifications and related engineering drawings. CDR should evaluate this documentation set and the detailed design to ensure that it adequately addresses the original requirements. Discrepancies raised during the PDR should be revisited to ensure that the rectifications are adequate. The review of TPM progress should be reviewed during CDR and all TPMs should be well within expected bounds. Particular attention should be paid to any TPMs that were highlighted during PDR as being potential problems.

- *Determination of readiness for production/construction.* In addition to the product specifications, ICDs, and the drawings, CDR should review the production plan and the quality assurance plans to ensure that the detailed design for the hardware items can progress to construction and production. The system prototype (if produced) provides an excellent tool for assessing readiness for production and construction.

- *Determination of maturity of software.* Software will need to enter the coding stage following CDR (this can be considered the software equivalent of hardware fabrication). The software product specifications (including interface requirements and design) should be thoroughly reviewed prior to approving the software aspects of CDR. It should be noted that current software development processes may dictate that software coding has begun long before construction and production. If this is the case, CDR should be used to monitor the progress of software development and investigate any software-related test and evaluation results.

- *Determination of design compatibility.* CDR should investigate and confirm design compatibility of CIs and other aspects of the system, such as facilities. This requires a detailed investigation of all external and internal interfaces. The system prototype will provide an indication of design compatibility.

- *Establishment of the product baseline.* The complete set of product specifications, once approved, will form the initial product baseline for the system, which is required prior to the conclusion of detailed design and development. Functional and physical configuration audits are eventually conducted on the production system prior to the formal approval of the product baseline for a system. These audits are discussed in Section 5.2.

4.6 Construction and Production

Once the product baseline has been established, following successful completion of detailed design and development, the system can move into construction and production.

Construction refers to the building of the system using the numerous components and subsystems defined in the preceding activities of the process. Production refers to the manufacturing or procuring of the components and subsystems that must come together to make up the system.

Some system development projects are aimed at producing only one system. There are many examples of one-off system developments, including high-rise building projects and some ship projects. Other system developments aim to produce many copies of the system. Our aircraft system is an example of a system where many system copies (aircraft) are likely to be produced in some form of production process. The initial aircraft (or even batch of aircraft) will be produced as prototypes to support design verification and user validation covered in more detailed in Section 5.3. Naturally, one-off or single systems have very different production requirements than do multiple systems.

Production requirements need to be considered early in the acquisition phase to ensure that production risks are identified and addressed as early as possible. As with all other technical requirements, the earlier that production issues are identified in the acquisition, the easier they can be addressed. To that end, although construction and production is being considered following detailed design, production engineers should be working with the rest of the design team from the earliest possible stages of the system development to ensure that production and construction issues are appropriately addressed. In short, the design that flows out of the preceding stages needs to be producible.

Typical construction and production issues that need to be addressed and monitored throughout the entire systems engineering process include the following:

- Material availability (lead times), ordering, and handling;

- Fabrication requirements, including tolerances;

- Processing and process control;

- Assembly, inspection, and test;

- Packaging, storage, and handling.

Strictly speaking, construction and production can be considered a specialty that supports the systems engineering processes. The systems engineering processes can be used before and during construction and production to support the construction and production activities. These processes include producibility analyses (ensuring that designs are in fact producible), consideration of producibility issues during trade-off studies and the selection of preferred design options, and LCC considerations that take the cost of production and construction into account.

Analysis-synthesis-evaluation is drawn upon again to support the construction and production effort. Figure 4.5 shows how this is applied to the construction and production effort.

The output of this process is a production plan that should be developed prior to the CDR and approved at that review. An approved production plan is required prior to entering construction and production. The content of the production plan varies widely with the specific characteristics of the system, but should include the following considerations:

- *Resources.* The plan needs to consider the type and size of any plant required to produce and construct the system. Personnel resources will include details of the skills required by personnel and any training programs required.

- *Production engineering considerations.* Production and construction schedules will be needed and the manufacturing methods and processes detailed. Special tooling and test equipment requirements must be specified. Any special facility-related requirements should be detailed, such as storage and layout requirements. If any automated

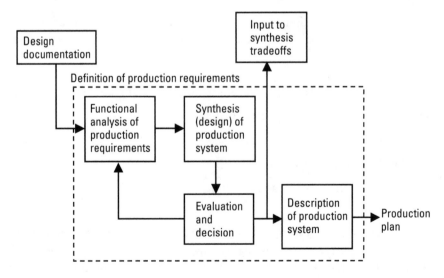

Figure 4.5 Systems engineering applied to planning for construction and production.

processes or manufacturing information systems are being used, these should be specified and described.

- *Materials and purchased parts.* A complete bill of materials should be compiled and listed. This will be available following the freezing of the detailed design documentation. COTS or modified COTS equipment should also be included in the plan so that all necessary items are covered. Any items involving long procurement lead times should be identified and ordered in a timely manner. An inventory control system is probably necessary to track all the parts, and this system should be specified.

- *Management.* The standard management functions should also be outlined in the production plan, which should include details on process control, production testing procedures, subcontractor control, and problem prevention.

- *Logistics.* The plan should also detail any special logistics requirements, such as packaging, handling, and storage requirements.

Before the system completes construction and production, it should be subjected to a range of acceptance tests and configuration audits. This is covered in more detail in Chapter 5. At this stage, it is sufficient to understand that the system must demonstrate the ability to satisfy the original user requirements that started the whole systems engineering process (acceptance testing). In addition to satisfying these requirements, audits should be conducted to ensure that the system has been built in accordance with the approved specifications and project documentation (configuration audits).

During construction and production, it may become clear that items of the system are not going to meet the all of the minimum requirements specified in the baselines. To cover this possibility, the configuration management process (see Chapter 5) uses a system of engineering changes, deviations, or waivers. Engineering changes are used where the departure from the specification is going to result in either a change in design or a change in requirement. If, however, the failure to meet the minimum requirements is temporary in nature, limited to only a small percentage of the supplies or a very minor departure, a deviation or waiver may be approved by the customer. Configuration management incorporating engineering changes, deviations, and waivers is a critical component of systems engineering management and is covered in detail in Chapter 5.

4.7   Operational Use and System Support

Once the system has passed the necessary testing and audits, it is ready to enter operational service or use—the utilization phase. The major activities during the utilization phase include operational use, system support, and modifications. The systems engineering influence over these activities is relatively minor and is normally confined to modifications, which should be made to the system in accordance with sound configuration-management procedures.

Configuration management must continue to play a role during modifications to the system to ensure that the system remains well documented as it moves through the utilization phase. Differences between system documentation and the physical system tend to result when systems are operated and maintained in an environment void of configuration-management practices. These differences make maintenance and operation difficult and potentially dangerous. Sound documentation supports the need to understand the design and capability of the system, especially if there are slight differences in configuration across a fleet of systems (for example, an aircraft fleet). Modifications made in accordance with configuration-management practices enhance supportability and operations, but unmanaged modifications adversely impact training, support, safety, and operations in the long run.

Modifications ensure that the system continues to meet operational and support requirements. Failures identified as part of the failure reporting, analysis, and corrective action system (FRACAS) process described in Section 4.7.1 may result in engineering changes to the system via the modification process. It may also be possible that modifications are required to rectify discrepancies in the performance of the system that were not identified during the acquisition phase. These discrepancies are often discovered during the operational test and evaluation effort when the system is placed in its operational environment and exercised in its intended purpose. Modifications may also become necessary due to changing system-level requirements caused by a range of factors, including a changing operational or support environment.

Regardless of their origins, modifications have the potential to significantly impact the system performance and functionality. Significant modifications can be considered systems in their own right, as demonstrated by Example 4.2.

Example 4.2: New Engine for Our Aircraft System Example

> *Adding a new engine to our aircraft system would seem to be a simple process of changing a few details and revising the manufacturing process to produce the new variant. A simple modification, when viewed with respect to*

its impact on the original WBS, can take on more significant proportions. Example 3.5 illustrated the concept of design space and the balancing act required to ensure all system-level requirements were met. Changing a major part of the system can significantly upset the function and performance of the overall system. In Example 3.3, the Engine CI illustrated how interface requirements are born. With these two examples in mind, a change in the engine has a large number of flow-on effects to other CIs and, ultimately, to system performance.

Consider the Aircraft System WBS in Figure 3.8 and the impact that a simple engine modification may have on all of the elements in that WBS. The air vehicle element will be changed significantly because the engine is a major component of the aircraft. It interfaces with a range of other CIs, so changing an engine will also require changes to other CIs. Test and evaluation on the engine itself and any system-level functions affected by the engine will need to be rewritten and performed. Training will be impacted as maintainers and operators will need to become familiar with the change. A new suite of technical and engineering data will be required to document the changes. Support equipment, spares, and facilities may also be impacted.

When the true scope of the modification is considered and investigated using the WBS, the modification starts to resemble a true system in its own right, and systems engineering methodologies can be employed successfully to achieve the modification.

Modifications have an impact on the original system; the degree of impact depends on the size and significance of the modification. Given that systems engineering can be employed to achieve modifications, systems engineering starts to play an increasingly important part in the utilization phase of the system life cycle as modifications become more prevalent. This gives rise to a revision of Figure 1.3 as shown in Figure 4.6.

System support also commences during the utilization phase to ensure that the system is maintained in operational order. The system support will be conducted in accordance with the approved logistics support concept (LSC). Modifications to the system may be necessary to ensure that the system continues to be supportable.

Example 4.3: Aircraft Modifications

Large aircraft systems such as military aircraft or airline aircraft are often designed to have a useful life of 30 to 50 years. During this time, changes in

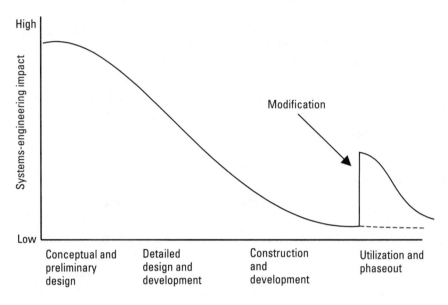

High

Low

Systems-engineering impact

Modification

Conceptual and
preliminary
design

Detailed
design and
development

Construction
and
development

Utilization and
phaseout

Figure 4.6 Effect of a modification on systems engineering impact.

technology areas such as engines and avionics will continue at a rapid pace. It is common for the operators of these aircraft systems to conduct major modifications to their fleets. Modifications may include replacing engines with more modern equivalents and replacing aging avionics with more sophisticated solutions, an example of which was described in Example 1.3.

Such modifications have a great impact on the aircraft system and may result in performance enhancements in the system, may allow additional tasks to be conducted using the aircraft, and may allow the aircraft to operate in a changed environment.

Although significant functionality enhancements often result from aircraft upgrade projects, the projects also help ensure that the aircraft system remains maintainable through to the aircraft's design life of type.

FRACAS

A FRACAS is a closed-loop system designed to continually maintain visibility into system operation and support. The FRACAS is put in place to record, analyze, and rectify the cause of system failures, especially recurring or related failures. The important difference between a maintenance system and a FRACAS is that the maintenance system rectifies the failure and the

FRACAS attempts to rectify the cause. A successful FRACAS improves the reliability and maintainability of a system as it proceeds through its life cycle as systems normally become less reliable and more difficult to support and maintain as they age. The FRACAS is in place to help counter this natural decline in reliability and maintainability.

The recurring theme throughout this text is that problems discovered early in the systems engineering process are generally the easiest, cheapest, and quickest problems to rectify. The FRACAS, therefore, should be established as soon as possible and becomes critical as the components, subsystems, and the system are produced, integrated, and tested. For COTS items, this could commence during preliminary design, and for developmental items, during detailed design and development.

A FRACAS based on MIL-STD-2155(AS) [3] has six steps as follows:

1. Failure reporting;
2. Failure analysis;
3. Failure verification;
4. Corrective action;
5. Failure report and closeout;
6. Identification and control of failed items.

An adaptation of this process is shown in Figure 4.7.

The system under investigation is operated or tested in accordance with the standard procedures until a failure is noted. At this stage, the failure is logged and data is collected to assist with the FRACAS process. The logging and data collection process needs to be documented, and all personnel involved with the test or operation of the system must be educated in the process to ensure that it is applied rigorously and consistently. Testing personnel are normally well versed in the FRACAS process, but sometimes failures go unreported during operations because operational personnel are not aware of the formal process in place. The success of the FRACAS relies on all failures being recorded, so it is critical that operational failures are included in the FRACAS—after all, the system spends a majority of its life in the utilization phase of the life cycle.

The data collected in the event of a failure needs to be sufficiently detailed to support the FRACAS process. The data needs to describe the conditions under which the failure occurred, symptoms of the failure, and quantitative descriptions of the result of the failure. The data supports the analysis process and also allows investigators to replicate the failure if necessary

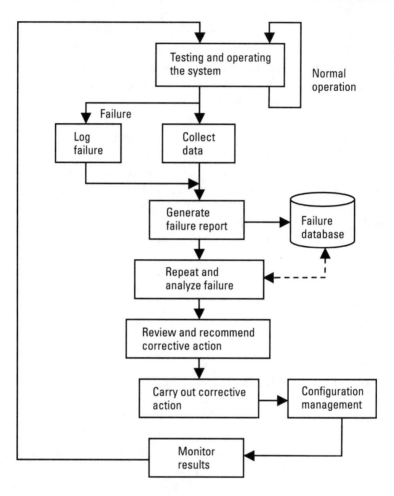

Figure 4.7 FRACAS process adapted from MIL-STD-2155(AS).

and practical. To that end, it makes sense to involve the investigators in the design of the form that will be used to collect the data. A failure report is generated to record the failure and its associated data. Modern FRACAS processes support this process with a database that simplifies the entry, storage, and retrieval of the information.

Once a failure report has been generated, responsibility for investigating the failure is delegated to the relevant personnel in accordance with the FRACAS process. The investigators are selected based on their experience and expertise and are delegated the necessary responsibility and authority to investigate the failure, analyze the failure, and generate a recommended corrective action report. These personnel investigate the failure using the reported data and searching the database for other similar or related failures. Under some circumstances, it may be necessary for the analysis process to repeat (and verify) the failure, but sometimes this may not be possible, especially if the failure results in injury or death of personnel or substantial loss of equipment. The analysis process makes use of established failure analysis tools, including failure mode effects and criticality assessments (FMECA), cause-and-effect analysis, fault-tree analysis, and so on. Following the analysis and verification process, the investigators determine the possible forms of corrective action likely to rectify the cause of the failures and generate a corrective action report containing their recommendations. Corrective actions are not limited to hardware or software changes, but can also include changes to the training of operators or maintenance personnel, or changes to operational or maintenance procedures.

The corrective action report is reviewed and approved by a review board established in accordance with the FRACAS process. The review board is closely associated with the established configuration-management process. If the recommended corrective actions involve class I or class II engineering changes, these are processed in accordance with the configuration-management process. The impact of the change is monitored to confirm that the failure does not recur as the system continues operation or test.

4.8 Phaseout and Disposal

Phaseout and disposal occur in the final stage in the system life cycle. Functions associated with phaseout and disposal include transportation and handling, decomposition, and processing. These sorts of issues should be addressed during the early stages of the acquisition phase of a system. If they are to be considered early, disposal and phaseout issues will need to form some of criteria against which the system is designed ("design for disposability"). This would ensure that attention was given to the issue early.

Although designing for disposability seems an unusual priority, consider the following examples of items that may be difficult and expensive to dispose of:

- *Radioactive material.* Although many projects would not be considered large users of radioactive material, common items, such as infrared detection systems, do contain radioactive material, and disposal raises a number of environmental issues.

- *Cryptographic material.* A large number of modern organizations make use of cryptographic material to protect information. Examples include the military, government, and financial institutions. The storage and use of the cryptographic material is well documented in various procedures, however the disposal of items such as hard disk drives and floppy disks that have contained this information is not a simple matter. Simply discarding the devices will leave the organization open to compromise, and a suitable disposal method must be formulated.

- *Normal building material.* Some materials considered normal have major impacts during disposal. Consider the environmental and health concerns surrounding asbestos. The disposal costs of buildings containing asbestos are extremely high. It is most likely that an alternative product would have been used in most buildings where asbestos is now a problem if its nature had been known during preliminary design.

Example 4.4: Disused Power Stations

In a nearby large city, we are aware of a disused power station located on prime real estate. The value of the real estate must be in the order of millions of dollars yet the site has remain untouched for many years.

One of the reasons for this is that the power station is full of asbestos, and the disposal costs associated with removing the power station are astronomical. It makes more economic sense for the owner of the land to leave the facility untouched, rather than dispose of the power station and sell the land.

In reality, phaseout and disposal are often not considered until they are imminent. Very little is written about formal methods for the inclusion of disposal and phaseout considerations in the early systems engineering activities. Perhaps as environmental and ecological constraints become more stringent, phaseout and disposal will begin to play a more prominent part in the system life cycle. Indeed, environmental considerations such as emission control and material recycling are already playing an important role in constraining the design, construction, operation, and disposal in such industries as the aircraft and motor vehicle sectors.

Endnotes

[1] *Systems Engineering Handbook—A "How To" Guide for All Engineers*, Seattle, WA: INCOSE, 1998, pp. 4.3–4.11.

[2] Further detail can be obtained in, for example:

Blanchard, B., *Logistics Engineering and Management*, Englewood Cliffs, NJ: Prentice Hall, 1992.

[3] MIL-STD-2155(AS), *Military Standard Failure Reporting, Analysis and Corrective Action Systems*, U.S. Department of Defense, Washington, D.C., July 24, 1985.

5

Systems Engineering Management

5.1 Introduction

The systems engineering management function is responsible for directing the systems engineering effort, monitoring and reporting that effort to the appropriate areas, and reviewing and auditing the effort at critical stages in the entire process. In this chapter we consider the major systems engineering management elements of technical reviews and audits, system test and evaluation, technical risk management, configuration management, the use of specifications and standards, integration management, and systems engineering management planning.

5.2 Technical Reviews and Audits

In this section, we consider the roles of reviews and audits in general, as well as providing further detail on the major reviews. Technical reviews and audits provide both the customer and contractor with a measure of progress toward the goal of successfully introducing a system into service, while reducing the technical risks associated with the system development. Reviews and audits achieve this by doing the following:

- Providing a formal evaluation of the design maturity;
- Measuring and reporting on planned and actual performance;
- Clarifying and prioritizing design requirements;

- Evaluating and establishing the system baselines at discrete points in the design process;
- Providing an effective means of formal communications between the stakeholders;
- Recording design decisions and rationales for later reference.

Technical reviews and audits are a vital part of the overall systems engineering effort. Reviews can range from the very formal and structured reviews normally associated with a critical system milestone to less formal reviews involving only a few people and a minor component of the overall system. Irrespective of the level of the review, all reviews aim to determine the ability of the design in question to meet the relevant technical requirements. Reviews tend to become more detailed as the design progresses and cover all aspects of the engineering design effort (for example, software, hardware, and integration). The customer normally specifies the required technical reviews and audits and their timing in the contractual documentation. The contract must specify the length, duration, location, content, conduct, and attendees for each review. Based on these requirements, the contractor is better equipped to estimate the cost and schedule associated with the reviews. A full set of reviews in a large project can represent a sizeable portion of both time and cost for the project. The contractor organization may also decide to include additional internal reviews, at their discretion, to further determine the technical adequacy of the design.

Obviously, the number of reviews and their respective scopes depend on the complexity and size of the system in question. The technical risk associated with the system also has an impact on the number of reviews scheduled. For example, a system that makes heavy use of unmodified COTS equipment may be seen as a low-risk project compared to a pure developmental project. Fewer, less stringent reviews would be expected for the COTS system than for the developmental project. In the case of modified COTS equipment, formal reviews may be limited in scope to cover the modifications.

It is clear that reviews and audits can provide significant assurance of the technical progress of the system development. These reviews and audits also take considerable time and effort to plan, prepare, and conduct. While reviews and audits are being conducted, attention is being diverted from the design and development effort. It is important that the costs and benefits associated with conducting technical reviews and audits are fully considered prior to planning and mandating the reviews and audits to ensure that an appropriate level of review and audit is placed on the system development effort.

Scheduling reviews and audits is also very important. Reviews held too early in the design process are unable to determine the technical adequacy of the design because the design is immature and subject to change. Reviews held too late in the process will miss the opportunity to avoid costly and time-consuming rectifications if the design proves inadequate. It is also useful to relate technical reviews and audits to documentation release [1]. For example, the PDR may be scheduled to coincide with the release of the hardware and software development specifications. Obviously, as the design progresses, the reviews and audits require hardware and software products as well as documentation, but the concept has merit because design documentation tends to be a strong indicator of design maturity.

This section introduces the major formal reviews and audits referred to in MIL-STD-499B [2]. Although this standard was never released, it still provides an excellent source of reference for formal technical reviews and audits and their intended purpose. As mentioned earlier, not all reviews are applicable to all systems. To that end, some reviews and audits can be tailored out of the requirements to suit the individual requirements of each system. Another excellent reference source for technical reviews and audits is MIL-STD-1521B [3], which was cancelled when MIL-STD-499B was released in draft. MIL-STD-1521B provides a detailed set of entry and exit criteria from each of the reviews and audits listed in this chapter.

Where system size and complexity warrants it, configuration audits may be considered as a separate topic from technical reviews. In this chapter, configuration audits are considered along with technical reviews. MIL-STD-973 [4] provides a comprehensive coverage of configuration audits, including the role that audits play in the overall configuration-management process.

5.2.1 Major Reviews

We have already considered a number of major reviews in Chapters 2, 3, and 4 when we considered the roles of SRR, SDR, PDR, and CDR as part of the system life cycle. Here we provide detail on other significant reviews.

5.2.1.1 Test Readiness Review

Testing is an expensive and time-consuming exercise involving highly trained personnel, expensive facilities, and specialized test equipment. The test readiness review (TRR) is sometimes contractually required by the customer to demonstrate that the system CIs are ready to enter formal test and evaluation. To that end, TRR occurs during construction and production and is designed to avoid the unnecessary expense involved with committing test and evaluation resources to a CI or system that is not sufficiently mature to enter testing.

To make an assessment about the readiness of the system to enter formal testing, the TRR normally reviews a range of documentation, including the following:

- *Test and evaluation master plan (TEMP).* The TEMP should be reviewed at each of the preceding design reviews but should also be rechecked at the TRR. Particular attention should be paid to the TEMP appendices, namely the test procedures to ensure that the procedures are correct and complete.

- *Formal and informal test results.* A major part of the TRR should be dedicated to reviewing the results of the formal and informal test and evaluation activities. These results give an excellent insight into the current status and performance of the CI. The test results should be documented in either the format specified in the TEMP or in some other approved way (for example, as documented in the contractor's quality assurance documentation). More credibility can be given to formal testing witnessed by either the customer or the relevant quality assurance organization.

- *Supporting documentation.* User and operator manuals are examples of documentation that should be reviewed as part of the TRR. Design documentation such as development and production specifications are also referred to during TRR. Customer test personnel most probably require this supporting documentation in the performance of the testing.

- *Support, test equipment, and facilities.* The test procedures list the support, test equipment, and facilities required to conduct testing. TRR should ensure that the equipment is available and in place prior to commencing test and evaluation.

5.2.1.2   Formal Qualification Review

The *formal qualification review* (FQR) may be required to verify that the performance of all of the CIs meets all of the functional requirements when integrated together into the system. The FQR will demonstrate that the integrated system complies with all of the specifications generated as part of the project including system specifications, hardware and software development specifications, and interface requirement specifications. From this point of view, the FQR could be considered a system-level configuration audit. For example, a formal qualification review may be required to verify that the software performs as designed when integrated with the actual hardware to be employed by the final system.

Normally, the FQR would be conducted following the functional configuration audits (FCAs) and prior to final system-level acceptance and operational testing of the system by the customer.

5.2.2 Major Audits

There are two main types of audit—the FCA and the physical configuration audit (PCA). Normally, FCA precedes PCA, but it is typical for both audits to be presented together.

5.2.2.1 Functional Configuration Audit

FCA is simply a check to ensure that the final as-built CI functionality as demonstrated by the test and evaluation effort is the same as the specified functionality for that CI in the relevant development and product specifications. The FCA may not be a discreet review of CI functionality; it may be conducted on a progressive basis throughout the development of the CI. Some CIs depend on the successful operation of other CIs and are not testable until after system integration. In these cases, the FCA is not complete until after the integrated system testing is complete.

All CI FCAs should be complete prior to releasing the design to full-scale production.

5.2.2.2 Physical Configuration Audit

PCAs are conducted on the final versions of the CIs (that is, the version of the CIs representative of the as-built items). PCA is an examination of the as-built CIs against the low-level specifications, such as the product specifications, assembly specifications, drawings, and technical data. The aim of the PCA is to confirm that the configuration of the CI as detailed in these low-level specifications is reflected in the physical as-built configuration. In the case of software, the product specifications and the code listings and software manuals are examined against the actual software CI to verify the same physical correlation. Once physical correlation has been confirmed for all CIs, the product baseline is said to exist.

By the very nature of the PCA, it is conducted during construction and production. The PCA is performed on the first production version of each of the CIs. In the case of a system where multiple copies are being produced, a partial PCA may be conducted on each production item to ensure that the production process remains constant.

5.2.3   Technical Review and Audit Management

The technical review and audit requirements of a project can be quite extensive, and the management of these requirements helps ensure that the overall aim of reducing technical risks and enhancing confidence in the design process is achieved. This section provides some management suggestions to assist with the technical review and audit requirements.

The requirements for reviews and audits must be specified clearly in the contractual documentation. The customer organization must consider the complexity and size of the system under development and tailor the review and audit requirements accordingly. Inadequate review and audit requirements may add risk to the system development by allowing potential design problems to go unchecked. The usual tendency, however, is to make the review and audit requirements bullet proof by over-specifying the requirements. Unnecessary reviews and audits increase the cost and schedule of the system without adding any additional value in the way of technical risk reduction. The additional cost and schedule slip are always ultimately passed onto the customer.

As well as specifying the reviews and audits required in the contract, the customer should also specify how the reviews and audits are to be conducted. The requirement for agenda, data packages, and minute taking will help in making the reviews and audits as effective as possible.

Both the contractor and the customer are active participants in reviews and audits. Both parties need to prepare prior to the event to ensure that the attendees have the appropriate expertise to contribute to the areas being covered in the review. The attendees must be given adequate time to review the agendas and data packages prior to the review. Each person should have a role in the review and participants should not consider the review as an educational experience. Observers may, however, attend the reviews as a convenient way for new staff to become familiar with the system/subsystem design.

Decisions, action items, and agreements should be noted in the minutes that are distributed following the meeting. Each action item should also be assigned to an individual who is responsible for following the action through to fruition by an agreed date. Unassigned action items invariably remain unactioned.

At the conclusion of the review, the customer team should meet to determine the level of satisfaction with the meeting. Successful technical reviews are a good indication of project status and schedule. Significant milestone payments are often tied to the successful completion of design reviews. Technical reviews and audits are, therefore, an important part of both the systems engineering and the project-management effort.

There must be a willingness to make difficult decisions in the event of unsatisfactory technical reviews and audits that may result in the need to revisit some or all of the preceding effort. The old adage that bad news does not improve with time certainly applies to inadequate design. While there may be pressure to meet progress and schedule targets, it is often better to stop work, go back a little in the process, and get it right before continuing. Succumbing to the pressure to continue is rarely in the interests of the project.

The technical review and audit effort is principally managed via a technical reviews and audits plan (TRAP). The TRAP documents all formal reviews, detailing the entry criteria that must be met prior to the commencement of the review or audit and the exit criteria that must be demonstrated prior to approval of the review or audit. The TRAP normally lists the design documents that need to be produced for evaluation during the relevant review. The TRAP is normally drafted during conceptual design and should be finalized and approved prior to any contract being signed.

5.3 System Test and Evaluation

System test and evaluation (T&E) is a systems engineering management function that ensures that a coordinated and consistent testing effort is applied to the system for the entire system life cycle. By coordinating the test and evaluation from a system perspective, the focus and emphasis of the testing can be varied with different life-cycle phases without compromising the entire test and evaluation effort. T&E applies throughout the entire system life cycle and involves both the customer and the contractor.

A thorough evaluation of a system involves validating the system against the original customer requirements. Obviously, this validation cannot be completed until the entire system has been designed, developed, and constructed, and then operated in the intended operational environment by operational personnel. The aim of system test and evaluation is to test and evaluate the system progressively as it passes through the acquisition and utilization phases in order to avoid costly and time-consuming modifications to the system design late in the life cycle. With this in mind, progressive test and evaluation could be considered a risk-mitigation measure, as it provides a high degree of confidence early in the system life cycle that the design will ultimately perform as required.

Testing can be an expensive and time-consuming exercise that requires the use of specialized personnel, test equipment, and facilities. A formal plan is normally put into place very early in the acquisition phase to manage the system test and evaluation effort. This plan is the TEMP, and it forms a major deliverable from the systems engineering effort. The TEMP is discussed in more detail later in this section.

The entire systems engineering process aims to produce a system that is both verified against the documentation produced during the systems engineering process and validated against the original needs, goals, and objectives that initiated the system development in the first case. Often these two associated aims are combined into the acronym *V&V*. V&V ensures not only that we have built the system right but we have also built the right system. A well-managed approach to test and evaluation aims to support the delivery of a system that is both verified and validated. Other systems engineering management functions, including technical reviews and configuration audits, also support the overall V&V objective.

There are three major categories of test and evaluation that are roughly applied to coincide with the acquisition phase, the transition between the acquisition and utilization phases, and the utilization phase:

1. *Developmental test and evaluation (DT&E).* DT&E refers to the test and evaluation activities undertaken during the acquisition phase of the system life cycle to support the design and development effort. DT&E activities may also occur during the utilization phase to support such activities as modification development.

2. *Acceptance test and evaluation (AT&E).* As DT&E completion approaches, AT&E activities become increasingly relevant. AT&E represents the formal acceptance testing conducted on the system to enable the customer to accept the system from the contractor. AT&E effectively forms the boundary or transition between the acquisition phase and the utilization phase. Unlike DT&E and operational test and evaluation (OT&E), AT&E tends to be a discrete testing activity (with a defined start and a defined end).

3. *OT&E.* OT&E is the term sometimes associated with the test and evaluation effort that is focused on the functional or operational testing of the system and its components, conducted under realistic operational conditions by operational personnel. OT&E is normally conducted for a period of time following acceptance of the system by the customer, although limited OT&E activities are possible during acquisition, especially where a long production cycle means that some systems have been accepted prior to other systems being produced, or when concept demonstrators are being used.

Figure 5.1 illustrates the role and timing of each type of testing.

This section describes the three forms of T&E and, where possible, details the specific test activities associated with each of the activities within the acquisition and utilization phases, as described in Chapters 2, 3, and 4.

5.3.1 Developmental Test and Evaluation

As the name suggests, DT&E occurs throughout the acquisition phase of the system life cycle and aims to highlight deficiencies in the system design as early as possible. As we have seen a number of times, the earlier problems are identified in the system life cycle, the cheaper and simpler they are to rectify. DT&E also provides confidence that the system, as designed, is likely to meet the original user requirements. The designers of the system also use DT&E results to validate the design process and minimize the risks associated with the system design.

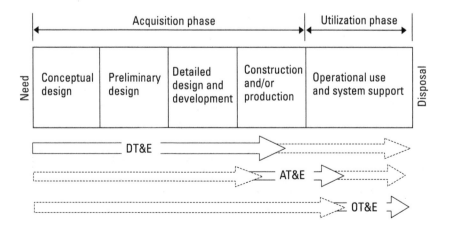

Figure 5.1 Role and timing of DT&E, AT&E, and OT&E.

DT&E covers a broad range of testing levels corresponding to the different design activities conducted during the acquisition phase. At one end of the spectrum, DT&E is responsible for test and evaluation of the lowest level components in the system. As such, DT&E effort may involve time on the test bench testing the initial hardware design mocked up on a breadboard. At the other end of the spectrum, full system prototypes (as described in Chapter 4) undergo test and evaluation as part of the DT&E program prior to the system being offered for AT&E and subsequent transition into the utilization phase. DT&E on system prototypes confirms that integration has been successful and provides confidence that the design meets requirements.

Responsibility for the planning and conduct of DT&E normally lies with the contractor. The customer organization normally places contractual requirements on the contracting organization with respect to test and evaluation, and it invariably wants visibility into DT&E. Visibility is normally provided during the design reviews, but it is not unusual for the customer to be invited to witness DT&E activities and be provided with copies of any DT&E reports and results. The customer may, in fact, consider accepting some DT&E results in lieu of repeating expensive and time-consuming testing procedures during AT&E.

DT&E tapers off as the system nears completion with initial delivery to the customer, however modification development during the utilization phase of the system life cycle revive the requirement for DT&E.

5.3.2 Acceptance Test and Evaluation

AT&E is a discrete activity normally shared between the contractor and the customer. The customer approves the specific acceptance test procedures to be completed as part of AT&E, but both the contractor and the customer may jointly conduct the tests. If the customer does not conduct the AT&E testing, they definitely witness the process and analyze the results. AT&E is focused on the system-level requirements contained in such documents as the system specification and the contract and is designed to provide confidence that the system meets the original user requirements as documented in the functional baseline.

The test results and any discrepancies are documented and rectified where required prior to the conclusion of AT&E. At the successful conclusion of AT&E, the customer accepts the system from the contractor, and the system formally enters the utilization phase of its life cycle.

Regression testing may be required where some acceptance test procedures have failed. The contractor will perform the necessary modifications to the system and the test procedures will be repeated. Care must be taken in these circumstances to understand what additional impact system modifications may have on other aspects of system performance. Regression testing should be conducted on all aspects of the system affected by the modification, which may involve significant effort.

5.3.3   Operational Test and Evaluation

OT&E is focused on operational functionality rather than design issues and is therefore commonly the responsibility of the customer organization. OT&E is normally conducted by an entity within the customer organization to assess the ability of the system and its components to meet the original user need under operational conditions. The testing entity is normally independent from the procuring agency within the customer organization to ensure that some level of independent assessment is provided. Separation of OT&E responsibilities from the customer's project office assists in maintaining objectivity in the assessment.

OT&E aims to estimate the operational effectiveness and suitability of the system. Potential modifications may be identified during the OT&E activities, and recommendations are passed through the appropriate channels to the designers. OT&E must be conducted in as realistic an environment as possible. Whereas DT&E can be performed in clinical conditions, OT&E must be conducted in operational conditions, including the use of real operators, support personnel, and logistics support.

OT&E may also be used to assist the customer in developing operational procedures for use of the system. It could be argued that a system continually undergoes OT&E throughout its life in service, albeit informally. Operations reveal shortfalls in performance and suggest areas for improvement.

Example 5.1: Testing in Our Aircraft Example

> *In our aircraft example, the company acquiring the aircraft (ACME Air) may be organized into three functional areas—operations, maintenance and support, and procurement. It is normally the responsibility of the procurement section to manage the project associated with the procurement of the new medium-range aircraft. To that end, procurement is responsible for customer T&E. Part of its role as the customer T&E agent is to involve operations personnel in the T&E effort, particularly in AT&E and OT&E. Through their T&E efforts, the company has much greater confidence as a customer in the new aircraft.*

5.3.4   Test Management

Regardless of whether the tests are conducted by the customer or the contractor, the contractor and customer should share responsible for managing DT&E, AT&E, and OT&E to ensure a coordinated approach to testing. Coordination will help maximize the effectiveness of testing and may also result in some savings in expensive testing resources, such as personnel, facilities, and test equipment. Management is necessary to ensure that the distinct testing functions are mutually beneficial, avoiding unnecessary duplication of effort and other conflicts.

5.3.5   Testing Activities and the System Life Cycle

Having introduced the three main categories of test and evaluation, it is instructive to investigate the focus of T&E activities in each of the stages within the acquisition phase.

5.3.5.1   Conceptual Design T&E Activities

Drafting the TEMP is a major T&E activity associated with conceptual design. From an engineering management perspective, the contracting organization drafts the TEMP to address the T&E requirements specified in the contract. As mentioned, the TEMP is also a contractual deliverable that details how the contractor intends to approach the overall test and evaluation effort during the acquisition phase of the system life cycle. The draft TEMP should be delivered as soon as is practical to the customer's project office for review and approval. It is unlikely that the TEMP is in its final approved form prior to completion of conceptual design. The very early versions of the TEMP produced during this phase of the life cycle lack the detail associated with a mature document, but the draft is still an important part of the T&E management effort. To that end, the TEMP continues to change (and be updated) throughout the acquisition phase of the life cycle as details become available. In some of the more popular TEMP formats, room is allocated for summaries of testing conducted to date. To that end, the TEMP is clearly a document that must be updated regularly.

DT&E helps the designers to select the most appropriate concepts and designs during conceptual design. Testing rarely occurs on any prototype systems, as these prototypes are not available until well into detailed design and development. An exception may be in the case where an existing system is being considered for modification and reuse to meet customer requirements. In this case, some system-level testing may be appropriate during conceptual design. The more typical DT&E conducted at this stage concentrates on analyzing key components of the different design alternatives under consideration to determine likely characteristics. In this way, selection of the preferred alternative is made easier. Feasibility studies considering different technology alternatives can also benefit from DT&E during this phase of the life cycle. With this in mind, it may be more appropriate to refer to the DT&E effort during conceptual design as analysis rather than evaluation.

The role of OT&E during conceptual design is very limited. Where similar systems exist elsewhere in the country or world, the customer may use these systems to develop a better understanding of their expectations and requirements. This is particularly valuable when the system being proposed is completely new, and the customer has no experience in its operation or capabilities. This idea of using a similar existing system to explore requirements and potential methods of employment is often called *concept demonstration* or *technical demonstration*.

5.3.5.2 Preliminary Design T&E Activities

The activities associated with preliminary design normally involve the validation of the concepts considered during conceptual design and development of draft testing plans and procedures. These plans and procedures are ultimately used to conduct and report on formal AT&E. The customer has an interest in these plans and procedures, and they are normally reviewed prior to the completion of preliminary design. Design validation will inevitably involve the development of some demonstration entities. For example, concept hardware may be mocked up on a breadboard arrangement, software modules may be produced using some form of rapid prototyping environment, and models may be constructed using some of the many design tools available today. These models and mockups do not necessarily represent the production equipment, but they provide a valuable vehicle for validating certain design concepts.

DT&E during preliminary design may include informal tests designed to exercise these models and mockups and collect the results. Designers are able to derive information regarding the system performance from these test results and evaluate the adequacy of the overall design concept under consideration (once again, the analysis-synthesis-evaluate loop). DT&E may also be required to confirm the performance of COTS or modified-COTS equipment selected for inclusion in the design.

As in conceptual design, there is little scope for OT&E during preliminary design, although in some development approaches, aspects of the system development may be mature enough to allow customer input. For example, designers of systems using very complex user interfaces may opt to design and develop the user interface completely prior to moving onto the back end of the system. The user interface may be advanced enough to be tested and accepted by the customer prior to preliminary and detailed design of the back end.

Most testing at this stage in the life cycle can be informal and is usually conducted in the "test bench" environment. Even though this testing is informal and concentrated on early engineering models, it is vitally important, as it provides the designers with the opportunity to identify any gross problems with the system design at a stage where design changes are relatively cheap and quick to implement.

In summary, T&E during preliminary design achieves the following aims:

- Evaluating any design areas that have been identified as significant risks in the overall design;

- Helping select the preferred technical approach to the system design from the alternatives under consideration;

- Helping to validate that the preferred technical approach has the ability to meet the original user requirements;

- Finalizing the T&E requirements for the remainder of the acquisition phase and document this in the TEMP.

5.3.5.3   Detailed Design and Development T&E Activities

By the latter stages of detailed design and development, the design has matured to an extent where more formal testing effort can be applied. System prototypes and mature software and hardware components are available for testing, allowing reasonably detailed and thorough testing to occur. The TPMs established and prioritized during conceptual design should provide specific testing requirements and expected results.

The DT&E effort during this phase of the acquisition concentrates on demonstrating that the design of the system is very close to complete and that all outstanding design issues have been adequately resolved. The testing must also provide some level of confidence that the design meets the requirements contained in the specifications. Examples of these requirements are provided next.

OT&E starts to be more focused at this stage in the process because the system is beginning to represent the final production system. Operational effectiveness is demonstrated through the OT&E effort at this stage in the process, provided the items being tested are sufficiently representative of the expected production models. Some development of operating procedures may also be possible during this testing.

Examples of some testing possible during this phase of the project include the following:

- *Functional testing,* which verifies that the specified functions can be performed by the system to the required level of performance under specified conditions.

- *Interface testing,* which ensures that the external and internal interfaces are operating correctly. This includes the system's ability to interoperate with other relevant systems and also involves verification of any human-machine interfaces.

- *Environmental testing,* which places the system in a range of environments to ensure acceptable operation under all specified conditions. Environmental testing will normally involve testing under conditions of varying temperature, humidity, terrain, and rainfall. Testing may also extend to cover less obvious environmental influences, such as electromagnetic effects and nuclear, biological, and chemical factors. The physical environment also must be investigated, including the effects of dust, vibration, and stress.

- *Physical and configurational testing,* which verifies that the system meets requirements relating to modularity, interchangeability, and accessibility. Tests also confirm that the physical characteristics of the system (such as size, mass, and volume) are acceptable.

- *Quality factor testing,* which includes early verification of the quality factors such as workmanship, reliability, maintainability, and availability.

Testing during detailed design and development is normally conducted at the supplier's facility. Problems and modifications identified during this testing are then easier to incorporate. The tests qualify the system to enter into construction and production.

There is no way of listing an exhaustive set of test categories that should be considered for a system, as the testing requirements of every system is completely dependent upon that system and its intended purpose. The main point to consider is that testing needs to be detailed and thorough enough to provide the necessary confidence to accept the system and transition it into the utilization phase.

5.3.5.4   Construction and Production T&E Activities

Once the system has successfully qualified through the tests conducted during the previous design stages, the system enters construction and production. This phase is focused on AT&E conducted in the later stages of this stage as systems begin to be produced. In most circumstances, AT&E is conducted by operational personnel from either the contractor or the customer. For example, by the time an aircraft project had moved through construction and production, production versions of the aircraft would be operating. Pilots from the customer organization would need to test the aircraft in as close an operational environment as possible. If the aircraft was a military attack aircraft, this environment may include an air weapons range including representative targets and threats. In a civilian context, other air traffic control radars and navigation aids would need to be considered as part of the environment.

Testing at this stage in the process may be the first time that all elements of the system have been fully integrated into a single entity. The indications provided by previous tests in areas such as quality factors can be verified and demonstrated throughout the testing at this stage of the development.

DT&E is now focused on the production process and procedures and on ensuring that production is operating correctly. The first system produced by the production process is tested fully. If more than one system is produced, each subsequent unit must be tested to an extent to ensure that the production and manufacturing process is continuing to be effective.

AT&E is focused on providing the customer with sufficient confidence in the as-produced system to accept the system as meeting the original requirements. AT&E can be as exhaustive as time and budget allows, but it is normal for some problems to surface after AT&E has been completed.

OT&E verifies that all production systems meet the original system requirements as they are accepted into service and recommends any modifications that may be required to the system. OT&E has the benefit of more time and a greater range of variables to stress and test the system than AT&E. OT&E eventually detects any problems with the production system.

5.3.5.5   System Utilization T&E Activities

Testing of the system continues throughout the utilization phase of the system life cycle. OT&E plays a relatively minor role in the T&E program up to this point in the life cycle. It is now possible for the fielded systems to undergo testing in a true operational environment. All components of the system are testing during OT&E, including facilities, training, data, logistics support, and the prime equipment. OT&E will be planned and conducted on an as-required basis but should be documented in the TEMP.

Typical T&E focus areas during the utilization phase include supportability issues, functional performance, and operation in the true operational environmental. This type of testing is conducted by the user organization and may result in modifications to the system. It also assists in the development of operating procedures and policies regarding the delivered system.

Example 5.2: A Training Package

The procurement of a complex aircraft system (such as in our example) includes delivery of a training system as part of the system. The training system is responsible for training air crew in the operation of the new aircraft, ground crew in handling procedures, and the maintenance practices necessary to keep the aircraft serviceable.

The effectiveness of the training system is only validated once all courses of instruction have been used on real students and those students have demonstrated the competencies required of trained personnel. This validation process takes place in the final training location, using delivered training aids and students from the customer's organization.

Validation of a training system may be partially achieved prior to system utilization, but sufficient confidence in the training system may only be provided following a number of successful courses conducted in the field. Validation of these courses is an example of OT&E.

DT&E may be required during the utilization phase to support such activities as modification development. As discussed in Chapter 4, modifications can be considered a system in their own right; therefore, DT&E is required during the acquisition phase of the modification process.

5.3.6 TEMP

The TEMP is the major planning document for the entire T&E effort planned for the acquisition phase of the project. The TEMP is also commonly referred to as the *master test plan* (MTP). The TEMP is required by the contract, prepared by the contractor, and reviewed and approved by the customer organization. The TEMP should be drafted prior to the end of conceptual design and reviewed as part of the technical review process. The TEMP should be in an approved state by PDR and should be reviewed at each of the major reviews until the completion of the acquisition phase.

The contract specifies the content of the TEMP but normally leaves the layout of the document up to the drafting organization. A typical TEMP may be required to contain the following information [5]:

- *System description.* This section includes a summary of the mission description of the system being developed and a description of the system that has been designed to meet the requirements. Major TPMs are listed in prioritized order to reinforce the mission description.

- *Program summary.* The program summary should detail the T&E responsibilities of all organizations participating in the T&E effort. The schedule of the T&E effort should also be outlined in this section of the TEMP to ensure that relevant organizations are aware of the proposed T&E activities.

- *T&E outline.* The TEMP should include reference to DT&E, AT&E, and OT&E in this section of the document. Critical technical characteristics (DT&E) and operational issues (OT&E) should be detailed, and any testing conducted to date should be recorded in the TEMP. As well as recording testing to date, the TEMP should detail the future testing (DT&E, AT&E, and OT&E). This way, the TEMP becomes a working document for the entire T&E effort.

- *T&E resource summary.* This section details the resources required for the entire T&E effort. Resources include personnel (both customer and contractor personnel), specific test and support equipment, test facilities, any simulators, models or test beds required, and any necessary personnel training required prior to testing. A schedule should also be included to show when the specific resources are required.

- *Appendices.* Any other documents relevant to the T&E effort should be attached to the TEMP. The set of test procedures that will be used to test the system should either be attached to the TEMP or referenced by the TEMP. Test procedures include the set-up procedure and initial conditions, a detailed set of instructions needed to run the test, and the expected results for each of the tests. The test reports generated following testing to record the results, determine whether or not the system has passed the test, and recommend a course of action for failed tests should also be either attached to the TEMP or referenced therein.

The TEMP may also reference other plans and documents (such as discrepancy reports) necessary to support the T&E process.

5.4 Technical Risk Management

Risk is defined as the possibility of a loss or injury or the possibility of some disadvantage or destruction. Systems engineering management is concerned with the management of technical risk or the risk associated with the technical aspects of the system life cycle. Broadly speaking, there are two major categories of risk: internal risk and external risk. Internal risks are those that are within the control of the project office; external risks are beyond the control of the project office. An example of an internal risk may be the risks associated with inappropriately manning the project office with qualified staff. This risk can be managed from within the project office. An external risk may be the introduction of some new legislation that places more stringent requirements on the system, leading to some changes to the user requirements.

The concept of risk is a subjective assessment in that risk identification and assessment are based on the perception of an individual and their interpretation of the term *risk*. What one individual may find risky, another individual may not. To that end, risk management must be performed by personnel with broad and experienced knowledge of the subject area to ensure that risks are assessed realistically. Given the extremely broad nature of risk, it is unlikely that a single person will be able to perform the entire risk-management function. Risk management is, therefore, a team effort. There are a number of components that combine to define the concept of risk. These risk components are listed here:

- *Risk identification.* The first step in the risk management process involves the identification of the potential risks relating to the system. Once the risks have been identified, they can be analyzed and managed accordingly.

- *Risk quantification.* Risk quantification consists of two parts: impact and probability. Risk impact estimates the likely damage that would be caused by the risk on the system should the risk actually occur.

Probability relates to the likelihood of the risk actually occurring. Clearly, the severity of any particular risk is a function of both impact and probability.

- *Risk response development and control.* The final component of risk management is the management (response development and control) of the risks identified and quantified by the previous steps. Management of the risks involves ranking the risks and devising a means to deal with the most critical of the risks. Classic means of dealing with risks include avoidance, mitigation, or acceptance.

Technical risk management is not a distinct systems engineering activity assigned to an individual risk manager, but rather an integrated part of a sound systems engineering management effort. This is demonstrated by the fact that a large number of tools and concepts described in this text as systems engineering processes or management functions contribute to technical risk management. Examples include the process of TPMs, technical reviews and audits, and configuration management (CM). Nor is technical risk management a one-off exercise conducted early in the acquisition phase of the system. Technical risk management is an ongoing concern. New risks may arise, existing risks may change, and old risks may disappear during the course of a system life cycle. To that end, technical risk management must continue throughout the system life cycle. Some popular standards such as MIL-STD-499B (reviewed in Chapter 6) contain guidance on risk management.

5.4.1   Risk Identification

Risk identification is the process of identifying the major areas of risk within a given system and documenting those risks. The process should document the source of risks and the symptoms that may be evidence of an impending risk occurring. Typical sources of risk in projects include changes in customer requirements leading to design changes, design errors or misunderstandings leading to unsatisfactory performance, and insufficiently skilled staff leading to longer design times or substandard designs.

No foolproof set of tools exists that guarantees the complete identification and documentation of all risks. Potentially the most effective means of risk identification is via a series of interviews or meetings with the appropriate groups of people. Given that both internal and external risks exist, the skill is in identifying the appropriate groups of people to interview.

Any source of information may provide insight into potential problems and risks. Sources of information include systems engineering design documentation, such as drawings and system architectures, trade-off studies, or design review documents. Other risk-management tools such as audit results, TPM monitoring, and T&E results also provide invaluable insight into potential new risks and the performance of existing risks.

It may be helpful to consider the following major categories as potential sources of risk during the identification process:

- Design, including design process, requirements, analysis, and testing;

- Testing, including the test reports containing test results;

- Production, including the manufacturing plan and process, test equipment requirements, and subcontractor controls;

- Logistics, including facility requirements, manpower and personnel, spares, and technical documentation;

- Management, including cost and schedule estimates, acquisition strategies, and staffing considerations.

By using these major categories, it may be simpler to assign responsibility for managing the identified risks.

Technical reviews and audits are also an excellent forum for discussing risk and potentially identifying new risks as the system moves through the acquisition phase. Strong consideration should be given to contractually requiring risk-management activities in each of the major formal reviews and audits.

5.4.2 Risk Quantification

Once the major risks have been identified, an analysis of each risk needs to be carried out to determine the likelihood of the risk occurring and the impact of that risk should it occur. This is called risk quantification. Based on the assessments of likelihood and impact, a severity of the risk (which is a function of both likelihood and impact) can be calculated. By attempting to analyze aspects such as likelihood and impact, the quantification exercise will focus on the cause and effect of the risk and in doing so provide some insight into the potential management of the risk.

A number of tools exist to assist in the analysis of risks, including classic project management scheduling tools such as PERT and critical path method

(CPM) charts, simulation, and models. A very popular tool, though, is expert judgment. There is benefit in using available tools to provide an input into the risk quantification, but it is easy to become overly focused on the answers given by these risk-quantification tools. It is vitally important to remember that risk quantification is primarily subjective in nature. There is a lot of merit in keeping the risk quantification relatively simple. Its main aim is, after all, to provide management with a list of potential risks (in order of severity) and provide some indication of the causes and effects of these risks. Management will then decide on the best available means of managing these risks as discussed in the following section.

A simple but effective means of ranking the severity of the risks is to categorize the likelihood of occurrence into three categories: highly likely, possible, or unlikely. Using the same process, the impact of the risk should also be categorized into three categories: high, moderate, or low. In this way, there will be nine possible levels of risk severity. To make the risk assessment practical, it is advisable to use less than nine levels of severity. Excessive levels add confusion without greatly adding to the ranking exercise. For example, it may be decided to use only three levels (high, medium, and low) as shown in Figure 5.2. By limiting the number of severity levels to three, the relative ranking of the risks is simplified.

5.4.3 Risk-Response Development and Control

There are three main options for the management of the risks identified and quantified in the previous two steps. These options are risk avoidance, risk mitigation, and risk acceptance. The management of risk should not focus solely on the negative aspects of risk, but also the benefits that may accrue from exposing the project to the risk.

		Likelihood		
		Highly likely	Possible	Unlikely
Impact	High	High	High	Medium
	Moderate	High	Medium	Low
	Low	Medium	Low	Low

Figure 5.2 Risk-severity table.

Risk avoidance is the option selected when the risk is too severe to be accepted at any level and must be avoided. Risks that may fall into this category are those that may result in loss of life or loss of business. The causes of the risk will need to be investigated to decide on how best to avoid the risk. It should be noted that by avoiding one risk, exposure to other risks may well result. This additional exposure also must be considered when deciding on how best to avoid a risk.

Risk mitigation may be the option selected when the risk falls in the middle ground in terms of severity. Mitigation basically means weakening, and, in the context of risk, mitigation means reducing severity. This can be achieved by reducing the impact of the risk, reducing the likelihood of its occurrence, or preferably both. Again the causes and effects of the risk must be considered in order to mitigate the risk successfully.

Risk acceptance is also a valid form of risk management usually reserved for risks of low severity. Risks may also be accepted when there is significant benefit to be gained should the risk not occur. The old adage of risk equals return does hold true sometimes for technical risks. For example, the use of leading-edge technology may be judged to be a reasonably severe risk, as the impact of it going wrong may be quite high. However, management may decide to go ahead with the use of the technology because if it does not go wrong, the benefits may be extremely attractive.

Risk-response development and control also considers how to monitor the risks identified. Severe risks should be monitored more closely and more often than less severe risks. However risks are managed, they must be managed continually to address the fact that circumstances are continually changing: old risks will disappear, new risks will appear, and existing risks will change in nature. Risk management is the responsibility of all parties involved in the system life cycle and must be performed continually.

5.4.4 Risk-Management Documentation

If there appears to be significant risk associated with a given system development, there may be a requirement for a risk-management plan (RMP), which is a formal document written by the contracting organization and reviewed and approved by the customer organization. The plan details the risk management process and the results of the application of that process for the particular development effort. In addition, the customer organization develops a more general RMP associated with all project-management-related aspects of system development.

The risk-management process should, as a minimum, contain the major steps outlined in this section, and the results (of the identification and quantification) will be dependent upon the particular nature of the system development. The RMP should detail the schedule for reviewing the risks and should contain details of the management (response and control) of each of the risks. The plan must be periodically reviewed because it is likely that risks will change with time.

5.5   Configuration Management

The term *configuration* can be defined as the relative disposition or arrangement of parts of something. In the context of systems engineering, the "something" is normally a configuration item that forms a part of the overall system and the "relative disposition or arrangement of parts" is called a *baseline*. To that end, CM is the act of controlling and managing the physical and functional makeup (defined in the baselines) of the configuration items that comprise the system [6].

The main aims of CM are to identify the functional and physical characteristics of selected system components, designated as configuration items, during the acquisition phase; to control changes to those characteristics; and to record and report on the change processing and implementation status [7].

Through an effective CM system, an accurate snapshot of the state of each CI (and through these, the system) can be provided. CM allows an effective history of changes to the system design to be maintained, including the alternatives considered and reason for selecting the preferred alternative.

For CM to be effective, the following functions must be performed throughout the project:

- Identification of the items to be placed under CM;

- Control of these items through an effective and documented change management process;

- Status accounting to provide accurate and up-to-date status information on the configuration of the items placed under control;

- Audits (such as PCA and FCA) to confirm the correct operation of the CM system.

5.5.1   Establishing the Baselines

The baselines used in the CM effort have already been introduced a number of times in Chapters 2, 3, and 4. Baselines are a very important part of system acquisition, as each established baseline provides a firm position from which to move forward.

- *Functional baseline.* The functional baseline of the system is defined by the approved system specification that is produced during the early stages of the acquisition phase. Once established, the functional baseline will be placed under configuration control, which means that changes to the baseline can only occur through a documented change-management process. This process is different for each project, but an example change-management process is described in this section.

- *Allocated baseline.* The allocated baseline for the system is defined by the set of development specifications for each of the CIs making up the system. These specifications should be developed and approved during the preliminary or detailed design activities of the acquisition phase. The allocated baseline is initially established when the set of development specifications is complete.

- *Product baseline.* The product baseline for the system is the set of product specifications (and any associated material and process specifications) for each of the CIs comprising the system. This baseline is established during detailed design and development and is approved by the successful completion of the configuration audits (FCA and PCA). Generally the functional baseline takes priority over the allocated baseline and the allocated baseline takes priority over the product baseline should any conflict or discrepancy occur.

5.5.2   Configuration-Management Functions

5.5.2.1   Configuration Identification

Configuration identification revolves around the selection of the configuration items during the early activities of the acquisition phase. Selection of CIs is a design decision documented through the breakdown of the system specification into lower level specifications, such as development and product specifications (type B and below). Typically, systems are broken into the major categories of computer software configuration items (CSCI) and hardware configuration items (HWCI). Naturally, there may be a number of CSCIs and HWCIs in the system design. Each CI is managed as a separate entity, including the design and test and evaluation of the item.

The selection of CIs is based on a number of factors:

- *Complexity.* One of the aims of the systems engineering process is to reduce the design problem to be manageable. Large complex CIs defeat this purpose, so CIs should be small enough to be able to be designed and developed by an individual or a small team.

- *Interfaces.* The difficulty with creating small CIs is that there will be a large number of them requiring more complicated interfaces. The selection of CIs is therefore a trade-off between the complexity of each CI and the number of interfaces required.

- *Use/function.* Each CI should perform a single major function or a combination of related functions.

- *Existing items available.* Rarely will a system design require all components to be designed and developed totally. Many of the components required by a system will already exist in a certain configuration, and the design should ensure that such components could be used rather than unnecessarily reinventing products. For example, if a reconnaissance aircraft has a functional requirement to record video, it makes sense for the functions to be allocated to a CI that looks something like a COTS video recorder.

- *Number of suppliers.* Similarly, unnecessary complexity is introduced into the design if a single CI is provided by more than one supplier.

- *Criticality.* A CI may be chosen to be smaller than would otherwise be required if the designers feel that the item is critical and therefore needs to be managed separately.

- *Maintenance and documentation needs.* Through-life support is an essential consideration in design, and CIs should be chosen to facilitate maintenance and documentation.

Once each CI has been identified, the configuration identification function ensures that the technical documentation required to completely define the physical and functional performance of that CI has been produced, and that those documents are current, approved, and available for use as required. The initial identification of CIs represents the translation from functional to physical design described in Chapter 3.

5.5.2.2 Configuration Control

At various stages throughout the system development, changes may be required. The reasons for these changes will be wide and varied, ranging from a change to one of the user requirements to a change in some available technology. Changes must be managed by a formal process to ensure that the impacts of the change are investigated and handled correctly.

Normally, a body called the *configuration control board* (CCB) is charged with the responsibility of managing changes. The CCB investigates the change and decides on the severity of the change. Major changes are sometimes called class I changes and include changes to form, fit, or function of the system, or major changes to system cost and schedule. As would be expected, the customer must be involved in the approval process of any class I change. Minor changes, or class II changes, include all other changes to the system configuration. Examples of class II changes include updates to documentation following identification of errors or changes to aspects of the design that do not impact on form, fit, or function, such as material substitutions. The customer should also be involved with class II changes but will probably be involved for information purposes only. The customer must, however, be involved in the process by which a change is classified as either class I or class II. The customer must be very aware of the potential ramifications of the change before agreeing to a classification of class II. This is because once a change is classified as a class II change, the customer can effectively lose visibility into, and control of, that change and its impacts.

The CCB should meet on a regular basis to review the progress of changes that are currently in progress and to review other proposed changes. Sometimes it is necessary to call a CCB meeting immediately to process a change that is considered urgent.

A change can be initiated by any of the involved parties, including the customer, the design team, or a subcontractor. In the case of a customer-initiated change, the customer may request that the contractor put together a change proposal for consideration by the CCB. This is requested because the customer is not in a position to provide the CCB with sufficiently detailed information regarding the impact of the change.

A change proposal must contain sufficient information for the CCB to decide whether or not to approve the change. The change proposal includes a statement outlining the change and the reasons that the change has been suggested. The proposal then details the alternatives available to implement the change and outlines the preferred alternative. The preferred alternative should be provided in sufficient detail to show that it will result in the change being implemented successfully and not introduce new problems into the system design. Of particular importance is the impact of the change on other CIs within the design. The impact of the change may result in changes to the interface requirements as specified in the baseline documentation. To that end, changes to the baseline documentation should also be included in the change proposal. The ability to assess impact relies heavily on sound traceability across the different levels of design.

In some cases, special interface teams are organized (comprising members from relevant CI development teams) to ensure that interfaces are handled correctly and completely. These interface teams will need to be involved with the change-management process, especially if changes to

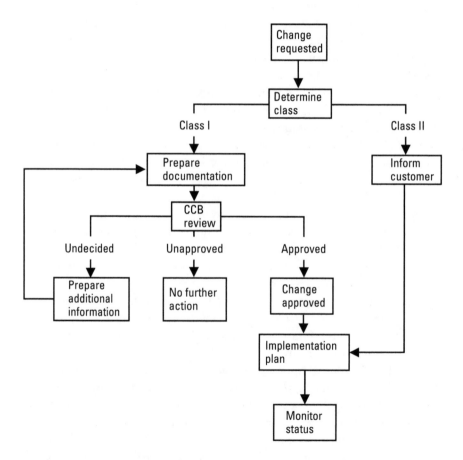

Figure 5.3 A simple change-management process.

the interfaces are likely. A simple change-management process is shown in Figure 5.3.

Deviations and Waivers

In some cases, it becomes clear during construction or production that items within the supplies are not going to comply with the relevant specification contained in the established configuration baseline. There may be any number of reasons for this occurring. Normally, these nonconformances would be rectified using the engineering change process just described; however, there are situations when departures from specifications may be accepted by the customer without the need to go through the engineering change process. Configuration management calls these special cases *deviations* or *waivers* [8].

A deviation is defined as follows [9]:

> A specific written authorization, granted prior to the manufacture of an item, to depart from a particular requirement(s) of an item's current approved configuration documentation for a specific number of units or a specified period of time. (A deviation differs from an engineering change in that an approved engineering change requires corresponding revision of the item's current approved configuration documentation, whereas a deviation does not.)

Example 5.3: Deviation

Following the analysis of a group of functional requirements, a design approach is developed that includes a computer system. The computer system is specified by the design team and is available as a COTS item. The COTS computer, as specified, is not available in time to support continuing integration, testing, and initial acceptance of the system. An option is to use a slightly inferior computer system that is available in the meantime. A deviation may be sought and approved prior to the procurement of the computer system to allow the designers to use the inferior computer system (even though it does not meet the minimum specified requirements) until the specified computer is available. In this example, the customer would be accepting a reduction in performance for a specified period of time as allowed for in the definition of a deviation. There is no need to seek an engineering change because the contractors are not seeking to change the configuration baseline of the system, nor is the change permanent.

A waiver is defined as follows [9]:

A written authorization to accept an item, which during manufacture, or after having been submitted for Government inspection or acceptance, is found to depart from specified requirements, but nevertheless is considered suitable for use "as is" or after repair by an approved method.

Example 5.4: Waiver

When the first aircraft is submitted to the customer for inspection, it is noted that the interior cabin space has a slightly different color scheme from that in the approved specifications. The customer has the option to reject the aircraft on this basis or approve a waiver from the color scheme requirements and accept the aircraft. In this case, the customer would probably approve a waiver and accept the aircraft for several reasons: it is a relatively minor departure from the specification, it can be rectified later if desired, and the benefits associated with accepting the aircraft outweigh the disadvantages associated with rejecting it. This example is a waiver because it has occurred after manufacture of the aircraft, and it represents a permanent departure from the requirements. The customer and the contractor should ensure that the cause of the departure is rectified so that the remainder of the production aircraft meet the color scheme requirements. There is no point in seeking an engineering change in this case because the system has already been produced, the customer does not want to change the color of the interior so a baseline change in not required, and the system is still deemed fit for purpose.

5.5.2.3 Status Accounting

Status accounting is the function that provides information about the current status of the items identified as requiring configuration control. This information includes the current state of the CIs and a history of any changes to the CIs from initial identification. Effective status accounting aids greatly in the change-management process, which in turn allows changes to be made to the various system baselines during the design process.

5.5.2.4 Configuration Audits

Configuration audits have been covered in Section 5.2 and include FCA and PCA. FCA confirms that all necessary functionality required of the CI is in fact included in the CI design. PCA concentrates on the physical configuration of the CI as described by the relevant documentation and displayed by the physical CI. If many copies of the CI are being produced as part of the system production process, usually the first production copy is audited. For items that are not physical in nature, such as CSCIs, the final versions of software listings are normally audited to ensure that the executable code is produced by a compilation of the documented source code.

5.5.3 Configuration-Management Documentation

A range of documentation may be required to support the configuration management effort for a system development. The documentation is normally specified in the contractual documentation. This section outlines some of the more typical pieces of documentation related to configuration management.

5.5.3.1 Configuration-Management Plan

The configuration-management plan (CMP) is one of the major systems engineering management plans and details the approach and processes involved in the configuration management for the project. There may be two CMPs for the system: The first is prepared by the contracting organization and applies to the acquisition phase of the life cycle, the second CMP applies to the configuration management of the system during the utilization phase. The second CMP is written and maintained by the customer organization.

The content and layout of the CMPs varies from system to system and is usually determined by the customer organization and dictated in the contractual documentation.

5.5.3.2 Change-Control Forms

There will need to be some basic form that can be used during the change-management process. This form is generically called the *change-control form*, which leads the CCB through the documented change-management process and will facilitate the initialization, evaluation, approval, and release of the changes as they occur. During the change process, the change-control form provides an excellent source of status information regarding the change.

5.5.3.3 Engineering Change Proposals

There are two major categories of change proposal relevant to the field of systems engineering management: engineering change proposals (ECPs) and contract change proposals (CCPs). As the names suggest, an ECP relates to the engineering design aspects of the project and the CCP relates to the contractual aspects of the project. ECPs and CCPs can exist independently of one another, but more often than not they will coexist. That is to say that an ECP will probably force a contract change (and therefore a CCP) and vice versa. This section concentrates on the ECP.

ECP content is determined by the customer organization and stated in the contractual document. As a minimum, the ECP should contain sufficient information to allow the CCB to evaluate and approve the ECP. This information will include a statement of the reason for the change, the various alternative solutions considered, and the preferred solution. Details of any trade-offs should be included, as should a statement of the impact if the change is not approved and incorporated. Another important piece of information that must be contained in the ECP is an impact assessment of the change that analyzes the impact the change will have on both other aspects of the design and the interfaces with other CIs, as well as the likely changes required to the current documentation set.

ECPs may also affect the current cost and schedule for the project—in which case, these impacts must also be included. It will be rare that a class-I ECP would be approved and incorporated with no cost and no impact on the schedule.

5.6 Specifications and Standards

One of the major outputs from the systems engineering processes is a complete set of documentation that accurately describes the designed system. It is not possible to understand the systems engineering processes without understanding how the documentation requirements fit into the overall system development effort. Every system development effort will have varying requirements with respect to the documentation set.

In addition to the documentation requirements, it is worth considering the role of standards in the overall design effort. These standards can take many forms, including military standards, international standards, and standards from recognized institutions, such as the Institute of Electrical and Electronic Engineers (IEEE) and the Electronic Industries Alliance (EIA). These standards bring together a wealth of knowledge and experience and provide the system designers with technical guidance during the system development.

Both standards and documentation requirements must be considered carefully by the customer organization before stipulating their use. Although documentation and the use of certain standards enhance the chances of project success, overdoing the requirements can add unnecessary overhead to the design effort. This overhead results in increased cost and longer acquisition schedules.

5.6.1 Specifications

Specifications are documents produced (or procured) during the acquisition process that describe the system or system components to varying levels of detail. There are five main categories of specification applicable to system acquisition projects and these are described briefly in this section.

5.6.1.1 System Specification

A system specification needs to state the requirements of the system in terms of both operational terms and maintenance and support terms. It must also document the interfaces between the system and its environment and between major functional areas of the system. Significant design constraints should also be detailed in the system specification. The requirements stated in the system specification should avoid reference to products and equipment and should concentrate on explaining the requirements in mission or capability terms. In its mature and approved state, the system specification will define the functional baseline of the system.

The system specification is also called a type A specification and is normally developed fully during conceptual design. The system specification is reviewed and approved at the SDR and, when approved, forms the functional baseline of the system.

5.6.1.2 Interface-Control Document

The ICD provides the mechanism for managing the integration of different system elements. The ICD is discussed in considerable detail in Section 5.7.

5.6.1.3 Development Specification

Development specifications are focused on subsystems (or CIs), as opposed to the system as a whole (the system consists of a number of subsystems). The development specification will detail the performance and interface requirements of each subsystem. The development specification may consist of two major parts: the requirements specification and the interface requirements specification. The development specification should be sufficiently detailed to allow designers to design and evaluate the subsystem design during the detailed design and development phase. An important requirement of the development specification is that it maintains a traceability of requirements back to the system specification.

The development specification is also called a type B specification. It is normally produced during preliminary design and will be reviewed and approved at PDR. It should also be noted that separate development specifications are written for hardware, software, and facilities (and any other major developmental component of the system). Standards such as MIL-STD-490A [10] define different types of development specification including prime item development specification and critical item development specification, depending on the item specified. The development specifications along with the system specification form the allocated baseline for the system.

5.6.1.4 Product Specification

The product specification is aimed at any subsystem or component below the system level that needs to be either procured (for COTS or modified-COTS) or designed and constructed. The product specifications for COTS or modified-COTS equipment are aimed at specifying the performance requirements for the equipment. Product specifications for equipment that needs to be constructed contain the detailed design requirements for that equipment flowing from the relevant development specification.

The contents of a product specification depend very much on the requirements detailed in the project contract; however, a typical product specification could be expected to include [11]: a complete set of performance requirements for the subsystem or component, a detailed description of the interface requirements, a detailed description of the parts and assemblies comprising the subsystem or component, and tests and expected results to verify correct fabrication of the subsystem or component.

For software products, an additional section may be added (if relevant) to detail any database-specific design and development requirements. If these requirements are insignificant, they may be included in the performance requirements section of the product specification.

The product specification is also called a type C specification and is normally produced during detailed design and development, although draft product specifications may be available as early as PDR. Mature product specifications are reviewed at CDR. Approval of the product specifications may be done in conjunction with the PCA. The approved product specifications for the entire system collectively form the subsystems or components baseline of the system. Product specifications should be produced for all subsystem or component within the system, including COTS, modified COTS, and developmental items. Obviously the documentation set delivered with COTS items will be less detailed than that produced for developmental items.

5.6.1.5 Process Specification

Process specifications are applicable to situations where a process must be performed on one of the subsystems or components within the system.

Examples of processes include painting, welding, soldering, or assembly. Normally, process specifications are written to define a specific or unique process to aid in the manufacturing process associated with the subsystems or components. Process specifications are also called type D specifications.

5.6.1.6 Material Specification

Material specifications are also written to support the production process and apply to raw material such as sheet metal, mixtures such as paints, and semi-fabricated materials such as fiber-optic cable. Material specifications are also known as type E specifications.

5.6.1.7 Specification Tree

Figure 5.4 shows a typical specification tree containing the specifications described in this chapter. The specification tree is used to illustrate how the lower level specifications flow from the system specification.

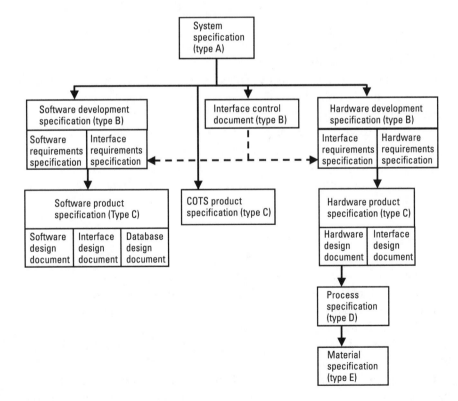

Figure 5.4 Typical specification tree.

5.6.2 Standards

An engineering standard is a generic term used to describe documents that establish engineering and technical requirements for processes and procedures that (when required by the contract) should be applied to a specific aspect of the acquisition phase. Standards have usually evolved over an extended period of time and are a refined source of significant expertise and experience. Even when standards are not contractually required, they should be considered as sources of information and potential guidance.

This section is aimed at outlining some engineering management issues relating to the use of standards during the system life cycle. Chapter 6 introduces the most relevant systems engineering standards in use today.

5.6.2.1 Overspecification

There is a general tendency by customers to overspecify requirements, including requirements relating to the use of engineering standards. The reason for overspecifying tends to be the belief that including more stringent requirements enhances the chances of success of the system development. More often than not, unnecessarily specifying the use of engineering standards introduces significant overhead without necessarily reducing the risks associated with the development. The increased overhead for the contractor is passed directly onto the customer in the form of extended schedules and increased costs.

The customer organization must be certain that the standards being specified in the contractual documentation are, in fact, required and not merely repeated there because "that standard has always been used in our contracts." This requires that the customer has read and understood the standard prior to specifying it.

5.6.2.2 Tailoring

A problem related to overspecification is that of the "blanket specification" of engineering standards. This means specifying the entire standard without consideration of the relevance (or lack thereof) of each of the sections within that standard. Standards are normally designed to be tailored to meet the specific requirements of each acquisition project. For example, certain technical reviews and audits may not be necessary in a particular project. If this is the case, the unnecessary audits and reviews should be tailored out of the standard that has been used to specify the review and audit requirements of the project. Without tailoring, unnecessary (and therefore expensive) activities and documentation may be undertaken by the contractor organization.

Tailoring should be completed before the tendering process begins for the acquisition project. This gives the tendering (and eventual contractor) organizations the ability to respond to the tender with accurate estimations of effort, cost, and schedule. In some circumstances, it may be acceptable to leave the tailoring until conceptual design and tailor the standards in a joint effort between the customer and the contractor organization.

Regardless of when or how the tailoring occurs, all specified standards must be considered in their entirety prior to the beginning of the acquisition phase. Unnecessary requirements must be tailored out before the requirements start needlessly impacting the acquisition effort. Effective tailoring can only be done by individuals who are sufficiently versed with the standard, the overall systems engineering discipline, and the specific system in question.

5.6.2.3 Unintentional Invocation of Standards

Standards sometimes reference other standards where necessary rather than repeating information that has been documented elsewhere. For example, standard A may reference the requirements in standards B, C, and D. Standard B may reference the requirements in standards E, F, and G, and so on. By specifying the use of standard A, the customer organization may also, by implication, unintentionally invoke the requirements of standards B through G. This implication inherent in some standards may result in an exponential explosion in the number of standards and requirements referred to in the contractual documentation.

The customer organization can avoid these problems by carefully selecting and tailoring the standards required by the contract.

5.7 Integration Management

In large and complicated system acquisitions, a wide range of technical specialists (or experts) will be involved with producing a wide variety of subsystems. The specialists perform work in the traditional engineering fields such as electrical, civil, mechanical, and aeronautical engineering, and in very specialized and detailed areas such as electromagnetic interference/compatibility and corrosion control. With all of these specialized areas working on the same project, there is a significant requirement for a coordinated approach to the integration effort. The integration effort must be managed to ensure that compatibility exists between the different areas. The task of systems integration falls within the bounds of systems engineering management and can be managed effectively by ensuring that adequate attention is paid to the functional interfaces between the different areas.

The task of integrating different system elements is managed primarily through the ICD, which was introduced in Chapter 3 as an example of a type B specification produced during preliminary design. The ICD is responsible for specifying the functional and physical interface requirements that exist between the different subsystem elements within the system. It should be remembered that interface requirements are defined in both the development and product specifications for the respective CIs. The ICD, however, identifies the CIs, between which there needs to be an interface. Through the definition of that interface, the ICD ensures that the communication occurs between the different design teams. This is a critical process because interfaces often result in additional derived requirements being placed on the subsystems. A formal interface document is a prerequisite for successful integration.

Example 5.5: Importance of the ICD

Suppose the system consisted of subsystem A and subsystem B. There would be separate development and product specifications for both A and B. The team responsible for A would not necessarily access the development and product specifications for B, and vice versa. A single ICD would be produced (called the A-B ICD or similar) defining the common interface requirements. Both teams would need to refer to this ICD because it will make reference to both subsystems. Both teams would also be responsible for cooperatively producing and maintaining the A-B ICD. It is clear that the ICD becomes the conduit through which the A and B teams communicate. In this way, the ICD assists in the systems integration effort.

The interfaces are defined in gradually increasing levels of detail as the design progresses from conceptual design, through preliminary design, and onto detailed design and development. Example 5.5 illustrates how design teams from different interfacing areas (that is software, hardware, and specialty engineering areas) must work as a team to bring the interface definition and design together. These teams are called ICWGs. ICWGs must meet regularly to generate the initial ICD and should then meet periodically or as the need arises to address any ongoing issues. These meetings and reviews may be informal, but the products of the meetings are formal interface design issues and decisions and updated ICDs.

ICDs are also formally reviewed by the customer organization as part of the formal technical reviews and audits process outlined in Section 5.2.

5.8 Systems Engineering Management Planning

Systems engineering is a broad subject; to cover the entire systems engineering effort, a systems engineering management plan (SEMP) is formulated to detail the required effort. The SEMP is normally constructed by the contracting organization in accordance with the requirements in the contractual documentation and reviewed and approved by the customer. Once approved, the SEMP becomes the governing plan controlling the entire systems engineering effort. Changes to the SEMP must be reviewed and approved as they occur, and the SEMP is normally reviewed at each of the formal design reviews.

The SEMP is fundamental when evaluating the ability of a contractor to perform the technical activities associated with the system development project.

Suggested contents for the SEMP are provided by a number of systems engineering standards. These standards detail the systems engineering requirements for systems development projects, and all emphasize the importance of the SEMP as the central systems engineering document. These standards view the SEMP as the tailoring of the standard to meet the specific systems engineering requirements of the particular project. The most popular of these standards are reviewed in Chapter 6. In addition, the U.S. Department of Defense provides a data item description for the SEMP [12].

The SEMP should cover all of the major systems engineering functions. It may do so by referring to other subordinate plans, such as the CMP, the RMP, the TRAP, the TEMP, specification lists/trees, and CI lists. Internal company design plans and processes should also be referenced by the SEMP if applicable. In this way, the SEMP completely defines the engineering management and processes to be applied to the project.

In addition to engineering management and processes, the SEMP should also detail positions of particular responsibility within the design team, including chief designers, software and hardware team leaders, project managers, and testing personnel. Part of the SEMP approval process should include an assessment of the skills and qualifications of these key personnel. Naturally, if this information is found in other management plans, the SEMP needs to refer to that plan.

The content of the SEMP should be maintained throughout the system design and development effort. Changes to the SEMP must be approved by the customer organization, as this ensures visibility into changes that may expose the project to unexpected risks (such as a change in key personnel).

Endnotes

[1]   Defense Systems Management College, *Systems Engineering Management Guide*, Washington, D.C.: U.S. Government Printing Office, 1990, p. 12-3.

[2]   MIL-STD-499B, *Military Standard—Systems Engineering—Draft*, Washington, D.C.: U.S. Department of Defense, 1994.

[3]   *Technical Reviews and Audits for Systems, Equipments, and Computer Software*, MIL-STD-1521B, 1985.

[4]   MIL-STD-973, *Military Standard—Configuration Management*, Washington, D.C.: U.S. Department of Defense, 1992.

[5]   Defense Systems Management College, *Systems Engineering Management Guide*, Washington, D.C.: U.S. Government Printing Office, 1990.

[6]   CM is a well-documented process. Some very useful sources of detailed information on traditional and modern CM practices can be found in the following:

      ANSI/EIA-649-1998, *EIA Standard—National Consensus Standard for Configuration Management*, Arlington, VA: Electronic Industries Association, 1998.

      MIL-HDBK-61A(SE), *Military Handbook—Configuration Management Guidance*, Washington, D.C.: U.S. Department of Defense, 2001.

      MIL-STD-973, *Military Standard—Configuration Management*, Washington, D.C.: U.S. Department of Defense, 1992.

[7]   Defense Systems Management College, *Systems Engineering Management Guide*, Washington, D.C.: U.S. Government Printing Office, 1990, pp. 11–1.

[8]   Other sources of information on deviations and waivers include the following:

      ANSI/EIA-649-1998, *EIA Standard—National Consensus Standard for Configuration Management*, Arlington, VA: Electronic Industries Association, 1998.

      MIL-HDBK-61A(SE), *Military Handbook—Configuration Management Guidance*, Washington, D.C.: U.S. Department of Defense, 2001.

      MIL-STD-481B, *Military Standard Configuration Control—Engineering Changes (Short Form), Deviations and Waivers*, U.S. Department of Defense, Washington, D.C., July 15, 1988.

[9]   MIL-STD-973, *Military Standard—Configuration Management*, Washington, D.C.: U.S. Department of Defense, 1992.

[10]   MIL-STD-490A, *Military Standard—Specification Practices*, Washington, D.C.: U.S. Department of Defense, 1985.

[11]   Defense Systems Management College, *Systems Engineering Management Guide*, Washington, D.C.: U.S. Government Printing Office, 1990, pp. 10–2.

[12]   DI-S-3618/S-152, Data Item Description, *Systems Engineering Management Plan (SEMP)*, U.S. Department of Defense, February 9, 1970.

6

Systems Engineering Management Tools

Systems engineering tools are available to assist both the systems engineering management function and the systems engineering processes, as well as assisting related disciplines such as project management. This chapter introduces a number of tools that may assist management (Chapter 7 introduces those that will assist with systems engineering processes). A number of standards and a capability maturity model are detailed and shown to assist the systems engineering management effort.

6.1 Standards

This section introduces some of the more prominent systems engineering standards and summarizes their contents. Repeating sections of text from the standards has been avoided, and readers should refer to the standard document if detailed content information is required.

The evolution of systems engineering standards is shown in Figure 6.1 to illustrate the history and previous standards upon which current standards are based. In the following sections, we briefly describe the standards from MIL-STD-499B onwards.

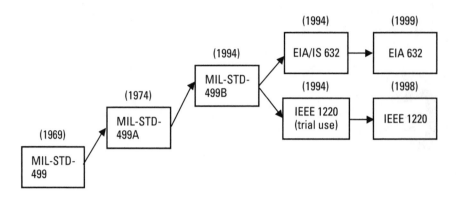

Figure 6.1 Evolution of systems engineering standards.

6.2   MIL-STD-499B Systems Engineering (Draft)

MIL-STD-499B was written within the U.S. DoD and released in draft form in 1994 to replace MIL-STD-499A, which was dated 1 May 1974. The Bravo version of 499 was never released other than in draft form. In 1994 the U.S. government, through the Perry initiative, directed that most U.S. military standards were to be cancelled. In fact, another standard reviewed in this section (EIA/IS 632) is considered to be the demilitarized version of MIL-STD-499B. Even though MIL-STD-499B was not officially released, it (like its predecessor) was, and still is, widely considered to be the premier systems engineering standard.

6.2.1   General Standard Content

MIL-STD-499B emphasizes that it focuses on explaining what to do rather than describing how to go about doing it. The standard attempts to assist with defining, performing, managing, and evaluating the systems engineering efforts required in each system development or acquisition project. The SEMP is seen as a central systems engineering document, and MIL-STD-499B recommends that both the customer (referred to as the *tasking activity*) and contractor (referred to as the *performing activity*) should produce SEMPs to describe their individual systems engineering efforts. The SEMP is seen as an effective way to tailor the requirements in MIL-STD-499B to meet the specific requirements of the individual project.

The standard has been written with a systems approach in mind in an attempt to make the standard relevant to a wide range of projects, from total systems development projects, through modifications and upgrades, to rectification projects. It is considered applicable to large or small projects, using new or existing technology, producing single or multiple units, being dominated by either hardware or software. Certainly MIL-STD-499B offers an excellent basis for all of those project types, but, again, the importance of tailoring the requirements needs to be emphasized. For example, the entire host of formal reviews and audits may be appropriate to a large and risky project but inappropriate to a small project making use of existing and mature technology.

MIL-STD-499B avoids the problems associated with the invocation of additional standards by avoiding references to other standards when specifying requirements. MIL-STD-499B uses different terms for the systems design and development stages of the acquisition phase. The stages used in MIL-STD-499B correspond well to the life-cycle stages used throughout this book in terms of activities and products. The stages used in MIL-STD-499B and the corresponding stages used in this text are shown in Table 6.1.

In the General Requirements section of the document, the standard calls for an integrated approach to the technical effort to best meet the customer's needs. This integrated effort must be reflected in the SEMP. MIL-STD-499B also requires schedule information to be contained in the SEMP and calls for two scheduling documents: the systems engineering master schedule (SEMS) and the systems engineering detailed schedule (SEDS). The SEMP (and the SEMS and SEDS) must be written in accordance with the tailored requirements of MIL-STD-499B, but once approved, the SEMP becomes the key systems engineering document for the project.

Table 6.1
Relationship Between MIL-STD-499B Life-Cycle Stages and Those Used in This Text

MIL-STD-499B Stage	Corresponding Stage in This Text
Concept exploration and definition	Conceptual design
Demonstration and validation	Preliminary design
Engineering and manufacturing development	Detailed design and development
Production and deployment	Construction and production
Operation and support	Operational use and system support

6.2.2 Systems Engineering Process

MIL-STD-499B sees systems engineering as having a series of inputs, which proceed through the systems engineering processes (with a management overlay) and result in a series of outputs. This is shown in the Figure 6.2.

The process starts with requirements analysis, which is aimed at taking the customer's needs, objectives, and requirements and translating them into functional and performance requirements for each of the systems primary functions. A functional analysis/allocation follows, which breaks the functions identified in the requirements analysis down to lower level functions. Traceability between low-level functions and the parent function is emphasized. Performance requirements should be defined for each high-level function so that measurable performance requirements can be allocated to each of the low-level functions. The systems engineering process chapters described a basic process of analysis-synthesis-evaluation. Requirements analysis and functional analysis/allocation collectively form the analysis stage of the basic process.

Next in the MIL-STD-499B process comes synthesis, which corresponds directly with our use of the term in this book. Basically, solutions must be designed and described for each logical group of functional and performance requirements previously defined. This design needs to be integrated to form a physical architecture. Following this design and integration task, a verification function must be performed to verify that the physical design does, in fact, meet the original functional and performance requirements. Verification forms part of the synthesis function in MIL-STD-499B but better describes the evaluation function described in Chapter 1.

Finally, MIL-STD-499B describes a process of analysis and control that can be described as a management overlay over the engineering process. Analysis and control is responsible for measuring progress, considering alternatives, and documenting the design decisions made. It includes functions such as CM, risk management, technical reviews, and the use of TPMs.

6.2.3 Content of the MIL-STD-499B SEMP

MIL-STD-499B talks of systems engineering planning as consisting of the SEMP, the SEMS, and SEDS. The SEMP is seen as the key planning document containing details of the intended systems engineering processes to be adopted. MIL-STD-499B also lists the following topics for inclusion in the SEMP:

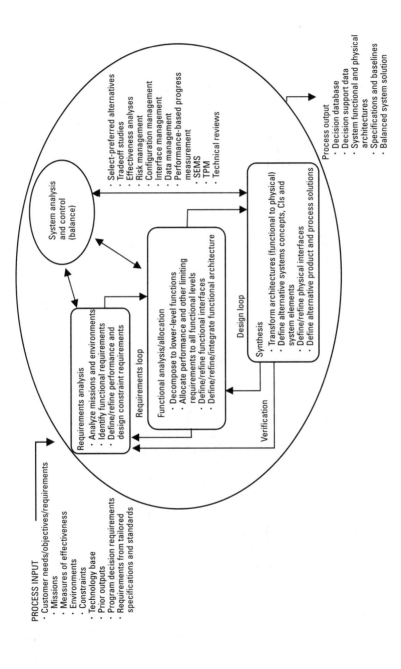

Figure 6.2 MIL-STD-499B systems engineering process [1].

- *TPM planning.* MIL-STD-499B states the importance of identifying technical performance measures that are "critical indicators of technical progress and achievement." The SEMP should detail how these are identified, how often they are updated and reviewed, the level of detail that TPM tracking covers, and the recovery planning process relating to TPMs.

- *Technical review planning.* The technical review planning should take into account both tasking activity reviews (formal reviews) and performing activity reviews (informal). All the tasks associated with each of the reviews should be specified, including responsibilities, venues, procedures, and entry/exit criteria. Although MIL-STD-499B does not make reference to a separate plan, it may be more appropriate to include these details in a separate plan called the TRAP and merely reference the TRAP from the SEMP. This may be particularly suitable if the review and audit process is complex and likely to change, in which case the TRAP can be updated without the need to update the SEMP also.

- *Technical integration planning.* The SEMP should detail how multidisciplinary teams and subcontractors will work together during the development effort, and how the many different products will be integrated. Again, an overview of the integration process with reference to other more detailed plans may be appropriate here.

- *Technology transition planning.* MIL-STD-499B makes reference to emerging technologies and the processes involved with transitioning this leading-edge technology from development to production and utilization. If this is applicable, MIL-STD-499B requires details of the transition planning in place to make the transition a success.

- *Relating TPM and SEMS/SEDS to schedule and cost performance.* MIL-STD-499B requires the performing activity to relate the risk-management tools of TPM and SEMS/SEDS to schedule and cost performance of the project. In this way, the information provided by the TPM and SEMS/SEDS can be more readily related to the project management aspects of the project.

6.2.4   Additional Information and Requirements

MIL-STD-499B lists and explains a number of functional tasks that need to be considered in a successful systems engineering effort. These tasks may be considered less obvious and may be potentially overlooked in the technical planning effort. The functional tasks listed in MIL-STD-499B are reliability and maintainability, survivability, safety and health, human factors, electromagnetic compatibility (EMC), security, producibility, test and evaluation, integrated logistics support, diagnostics, transportability, and infrastructure.

All of these tasks must be considered during the technical planning process. In some projects, all areas may be relevant, but in the majority of cases some of the tasks will not be applicable. Each project must be considered individually as emphasized during the discussions on standards tailoring.

The concept of leverage options is introduced, which relates to design options for reducing the risks and costs associated with design. Performing activities must consider options such as the use of nondevelopmental items (NDI), dual-use technology, open system architectures (OSA), and re-use during the planning stages. Dual-use technology is defined as "technology that can be used in either the commercial or government sectors or both" [2]. OSA basically means use of architectures that are widely used and have become industry standards. By requiring the use of OSAs, MIL-STD-499B is avoiding proprietary solutions that may become obsolete and difficult and expensive to support. These options should be documented in the SEMP.

MIL-STD-499B highlights (in a section entitled "Pervasive Development Considerations") some related support issues that may need addressing in the SEMP.

The computer resources necessary to support the design and development effort must be considered and planned for. Issues such as the processor capabilities, development tools, and development environments must be considered to ensure that the hardware is capable of supporting the mature version of the system software.

Parts, materials, and processes needed to support the development effort are critical and also need planning. This section must address inventory and supply support issues and must take LCC into consideration. Intended use of simulation and prototyping to support the design process should also be detailed in the SEMP.

The use of digital data, computer-aided design tools and other design technologies needs to be specified. This requirement may be showing the age of MIL-STD-499B, as these techniques were only emerging in the early 1980s when the standard was initially drafted whereas computer-aided design is commonplace today. Even so, there may be a requirement for the performing activity to include these details in the SEMP.

MIL-STD-499B emphasizes that cost effectiveness should be considered throughout the project. The standard requires that a range of analyses be considered and documented by the performing activity with a view to improving the cost effectiveness of the project. The analysis tasks listed in MIL-STD-499B include manufacturing analysis, verification analysis, deployment analysis, environmental analysis, supportability analysis, training analysis, LCC analysis, operational analysis, and disposal analysis.

A large section of MIL-STD-499B is focused on technical reviews and audits. The section provides a summary of the formal reviews and audits that can be conducted during development, and it provides details such as when the reviews and audits should be conducted and what should be achieved from the reviews. The review responsibilities are also listed in terms of tasking and performing activities. The most critical (and commonly used) reviews and audits from MIL-STD-499B have been listed and described in the appropriate section in Chapter 5.

MIL-STD-499B also provides a section of notes, which contains some very important tailoring and application guidance covering the use of the standard in different situations. The importance of thorough and thoughtful tailoring has been emphasized throughout this text. The notes section also contains a list of suggested military standards for related topics such as configuration management and software engineering. This list is out of date and contains standards that are no longer used or supported, but the list does provide an interesting summary of topics related to systems engineering management.

There are also three appendixes to MIL-STD-499B, namely a glossary of terms, a list of acronyms and abbreviations, and some guidance on the development of the SEMS.

6.2.5   Summary

MIL-STD-499B is out of date and was never issued or supported by the U.S. DoD. Despite this, the standard is an excellent source of information and guidance on the subject of systems engineering management and technical planning. The relevance of MIL-STD-499B is demonstrated by the almost word-for-word adoption by the EIA in their interim systems engineering standard (EIA/IS-632).

The contents of MIL-STD-499B reflect a vast and varied range of experience in the field and provide both customer and contractor organizations with learned advice on engineering management. MIL-STD-499B remains a highly recommended source of systems engineering information.

6.3 EIA/IS-632 Systems Engineering

In 1994, it became evident that MIL-STD-499B was not going to be released by the U.S. DoD as the military standard on systems engineering. The EIA had an established committee concentrating on systems engineering, and that committee undertook the task of adapting MIL-STD-499B for use as an industry standard. That adapted standard is called EIA Interim Standard (IS) 632 and was released in December 1994. Work on the updated standard was recently completed in 1999 with the release of ANSI/EIA-632.

A number of organizations were involved with EIA during the development of EIA/IS-632, including the U.S. DoD, the National Council on Systems Engineering, American National Standards Institute (ANSI), and the IEEE.

6.3.1 General Standard Content

The result of the work carried out by EIA during the development of EIA/IS-632 is an industry standard that represents a demilitarized version of MIL-STD-499B. The differences in general content between MIL-STD-499B and EIA/IS-632 are minor. By scanning the respective tables of content, the similarities become obvious.

6.3.2 Systems Engineering Process

The systems engineering processes detailed in EIA/IS-632 are identical to those contained in MIL-STD-499B. EIA/IS-632 has added some additional requirements in the requirements analysis and functional analysis and allocation areas. The additional requirements involve a verification function as part of each stage to ensure traceability of both requirements and functions to either systems engineering inputs (customer requirements and needs) or to a higher level requirement identified earlier in the process. Note than both EIA/IS-632 and MIL-STD-499B require verification as part of the synthesis process; however, MIL-STD-499B does not mention verification as part of requirements analysis or functional analysis and allocation.

The addition of the verification function in all of the engineering processes adds a degree of rigor to the requirements contained in EIA/IS-632 and ensures adequate traceability in all of the systems engineering processes.

6.3.3   Content of the EIA/IS-632 SEMP

The content of the SEMP as required by EIA/IS-632 is largely the same as the MIL-STD-499B SEMP. Recall that MIL-STD-499B required the SEMP, SEMS, and SEDS to form the basis of the systems engineering planning effort. To this list, EIA/IS-632 additionally requires the WBS to play a role in the planning process.

6.3.4   Other Information and Requirements

The remaining information and requirements contained in EIA/IS-632 are identical to those contained in MIL-STD-499B. The review and audits section of both standards refers to same set of formal and informal reviews and provides the same guidance on issues such as roles, responsibilities, expected outcomes, and entry/exit criteria.

EIA/IS-632 contains a notes section (as in MIL-STD-499B) that includes important information on the tailoring and application of the standard for different projects.

The appendixes in EIA/IS-632 are slightly different from those in MIL-STD-499B in that EIA/IS-632 provides an appendix detailing the expected structure and content of the SEMP, including details such as preparation and numbering conventions. This appendix provides useful guidance and will help to ensure uniform standard and content across SEMPs formulated by different organizations and different projects. EIA/IS-632 also provides an additional useful appendix containing information about the SEDS, which is not contained in MIL-STD-499B.

6.3.5   Summary

EIA/IS-632 is almost identical to MIL-STD-499B in all respects. Some military-specific phrases and examples have been altered to make the EIA/IS-632 standard less military focused. These changes are merely cosmetic in nature. The other changes (as detailed in this section) are very minor and do not make EIA/IS-632 substantially different from MIL-STD-499B.

As with MIL-STD-499B, EIA/IS-632 contains excellent information, advice, and guidance. It is worth remembering that MIL-STD-499B was only a draft and was never officially released for use by the U.S. DoD. The major advantage, therefore, that EIA/IS-632 has over MIL-STD-499B is that EIA/IS-632 was officially released by the EIA. For this reason, the use of EIA/IS-632 over MIL-STD-499B is recommended.

6.4　IEEE 1220 (Trial Use) and IEEE 1220-IEEE Standard for Application and Management of the Systems Engineering Process

The IEEE started work on IEEE 1220 in 1989. Early work on the standard focused on parallel effort with the development of MIL-STD-499B to ensure consistency between industry and military/government systems engineering direction. When released in 1994, IEEE 1220 was the first commercial systems engineering standard. Because of this, IEEE 1220 was released as a trial-use standard. Since the initial release of IEEE 1220 in 1994, EIA/IS-632 has also been released. IEEE updated the trial use standard into a full standard in December 1998 but retained the designation IEEE 1220. There are only minor differences between the 1994 and 1998 IEEE 1220 standards.

IEEE 1220 has the same scope as defined in both MIL-STD-499 and EIA/IS-632 in that it is designed to be applied to the technical effort of a wide range of projects covering development, modification, or rectification of equipment. It is applicable to large or small projects, using new or existing technology, producing single or multiple units, being dominated by either hardware or software. IEEE 1220 has been written to be applicable to commercial, government, or military projects.

The systems engineering planning requirements contained in IEEE 1220 center around the engineering plan, which is the result of the tailored application of the standard. Tailoring is again emphasized as an important part of the application of the standard to different situations and projects.

IEEE 1220 is focused on a system paradigm in that everything within the system (including products, subsystems, and processes) can be dealt with as a smaller system within the main system. This is possible because products, subsystems, and processes all have boundaries, relationships, and interfaces with the other products, subsystems, and processes within the main system. In this way, the requirements of IEEE 1220 can be applied to each individual system within the main system, ensuring a consistent approach to the technical effort behind the project.

6.4.1　General Standard Content

IEEE 1220 presents a very familiar systems engineering process. The process is consistent with the processes adopted by MIL-STD-499B and EIA/IS-632 and detailed in this book. The process is applied at each stage of the system life cycle and at each level of system decomposition. The IEEE 1220 process is summarized in Figure 6.3 [3].

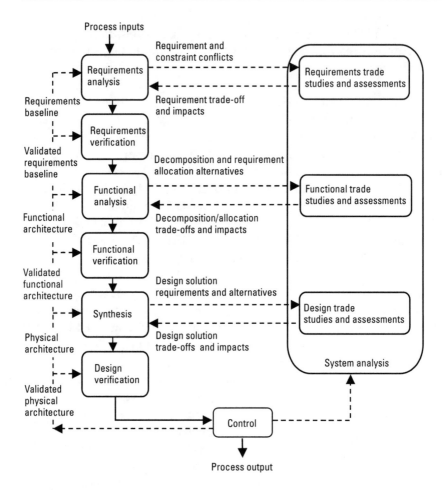

Figure 6.3 IEEE 1220 systems engineering process [3].

The first step in the systems engineering process is the analysis stage and this is represented in IEEE 1220 as the requirements analysis and the functional analysis. The synthesis function exists in both IEEE 1220 process model and the systems engineering process model used in this text. The final stage of the systems engineering process is evaluation. IEEE 1220 emphasizes the evaluation function throughout the entire process. Accordingly, evaluation functions occur at each stage in the process and are called requirements baseline validation, functional verification, and physical verification.

The engineering plan plays a central role in the technical planning when using IEEE 1220. In addition to the engineering plan, the master schedule, detailed schedule, and various technical plans are also referenced. IEEE 1220 explains that typical plans produced to support the engineering plan include risk management, technical reviews, configuration management, verifications, computer resources, manufacturing, maintenance, training, security, and human systems engineering.

The importance of design data is emphasized in IEEE 1220 through the requirement for an integrated database to capture and track design data. An integrated data package is also required to document the architecture and design information, as necessary to support manufacturing and through-life support. IEEE 1220 lists and briefly describes the specifications and drawings that are expected in a typical integrated data package.

IEEE 1220 refers to a tool called the *system breakdown structure*, which takes on the same role in IEEE 1220 as the WBS in the other standards. The roles and importance of quality assurance, systems integration, and technical reviews are emphasized in the "General Requirements" section of the document.

6.4.2   IEEE 1220 Life-Cycle Model

IEEE 1220 dedicates an entire section to the application of systems engineering across the different phases of the system life cycle and provides a comprehensive description of the activities within each stage of the life cycle.

The IEEE 1220 life-cycle stages (and the corresponding stages in this text) are described in the following sections:

- *System development.* System development in the IEEE 1220 life-cycle model best corresponds to conceptual design in the model in this text. This phase of the life cycle (according to IEEE 1220) concentrates on the conceptual design of the system, the development of project and technical plans, consideration of major interfaces (including human/system interfaces), identification of major risk, and the development of system-level specifications. The system baseline, which corresponds with the more generally used term, *functional baseline,* will result from the system development effort. A review concludes the phase and selects the preferred concept and ensures that the system-level requirements are completely defined.

- *Subsystem definition.* System definition in IEEE 1220 is the major development phase of the system life cycle and consists of a number of subphases.

- *Preliminary design.* The IEEE 1220 preliminary design activity corresponds directly to preliminary design as presented here. Subsystem identification, definition, and design are considered during this stage of the development, including interfaces, risks, and specifications. The project and technical planning documentation will also be reviewed and updated where necessary. The result of the preliminary design phase is a design-to baseline, which corresponds to the more generally accepted allocated baseline. The phase is concluded with the conducting of a design review to ensure that the preliminary design is complete and satisfactory.

- *Detailed design.* The detailed design phase of the IEEE 1220 life cycle corresponds to the detailed design and development phase as presented here. In this phase of the development, the subsystem design is further broken down to the lowest component level, thus completely defining the subsystem design. Component definitions and risks are documented in the form of detailed drawings, parts lists, and specifications. The work in this phase of the development results in the build-to baseline that we have called the product baseline. Technical reviews are focused on ensuring component-level definition and design is satisfactory. The design reviews completed following detailed design may also revise the system and subsystem designs if any changes have occurred.

- *Fabrication, assembly, integration, and testing (FAIT).* FAIT activities are the beginning of the construction and production phase. System integration and testing occurs in this phase where the components and subsystems are integrated and analyzed to ensure that the higher level functionality and interfaces are satisfied. Where problems are identified, these are fixed and retested. The main reviews conducted during this phase include the TRR and the FCA. Successful completion of this phase provides approval to move the system into production.

- *Production and customer support.* This final phase of the IEEE 1220 life-cycle model encompasses both construction and production and the utilization phase. Production-related activities such as inventories, production control, and integration occur. A PCA is conducted on the initial production system to ensure compliance with the technical documentation. Customer support and after-market products are provided as required.

6.4.3   Systems Engineering Process

IEEE 1220 details the tasks that need to be performed in each of the process steps shown in Figure 6.3 in an analysis-synthesis-evaluation loop. Each of the processes is summarized using a flow diagram in the standard.

- *Requirements analysis.* The requirement analysis activity is designed to establish what the system is required to accomplish in quantitative and measurable terms. The analysis must also define the environmental constraints within which the system will be required to operate. This information forms the requirements baseline upon which the remainder of the development effort will be based. Sixteen subclauses are detailed in IEEE 1220 pertaining to conducting the requirements analysis effort. These subclauses provide direction and guidance to those responsible for performing the requirements analysis effort. Clause 6.1.16 describes the requirements baseline, which is the product of the analysis and is recorded with three views: operational, functional, and design. The operational view is written from a user's perspective and describes what the system must be capable of and how it will be used. The functional view describes what the system products must do to meet the operational requirements. The design view concentrates on the physical interfaces, the human engineering aspects, and the physical form of the system.

- *Requirements validation.* Requirements validation concentrates on the requirements baseline produced during the requirements analysis. It ensures that the baseline represents the user's requirements within the identified constraints and that the baseline is a complete representation of those requirements. Five subclauses are detailed in IEEE 1220 to describe the validation process. There may be a requirement to revise and update the requirements baseline during the validation process if deficiencies are identified.

- *Functional analysis.* Functional analysis takes the requirements in the requirements baseline and further defines the problem in clearer detail. This is done by translating each requirement into functional terms and into a functional architecture. Decomposition of the requirement into functions and subfunctions (including interfaces and performance requirements) continues until the description is complete. Functional analysis does not result in a design solution. Grouping of logical subfunctions will occur during the synthesis stage to produce a design solution. IEEE 1220 provides eight sub-clauses to direct the analysis effort.

- *Functional verification.* Functional verification investigates both the completeness of the functional architecture in line with the requirements baseline and the suitability of the architecture as an input into the synthesis stage. The importance of upward traceability from the functional architecture to the system-level requirements and constraints is emphasized. Also, downward traceability from the system level to the functional level will help ensure that all requirements have been addressed.

- *Synthesis.* Synthesis defines the product solutions and subsystems that will meet the functional architecture requirements. First, the functions are grouped together and a range of design solutions considered. The most suitable of these alternatives is selected with associated documentation verifying the selection. A range of considerations is used during this selection. Synthesis results in a design solution defining integration arrangements of subsystems and the internal and external interfaces. IEEE 1220 provides 18 subclauses to direct the synthesis effort.

- *Design verification.* Physical verification is primarily aimed at ensuring traceability from the lowest level physical architecture to the functional architecture verified during the functional verification stage. This traceability should also be continued through to the requirements baseline. There are a number of verification approaches possible, including inspection, analysis, and demonstration. The method of verification will need to be defined prior to the exercise.

- *Management layer.* The management layer identified in IEEE 1220 consists of systems analysis and control. Systems analysis includes activities such as trade-off studies to consider alternatives, risk assessments, and LCC analysis. Control covers such issues as design data management, technical management, configuration management, risk management, interface management, and project and technical planning.

6.4.4   Content of the IEEE 1220 Engineering Plan

IEEE 1220 provides a very useful systems engineering management plan template as Annex B to the standard. The rest of the standard refers to this plan only as an engineering plan, not a SEMP. The content of the IEEE

1220 engineering plan is designed to allow the tailoring of the standard to suit the project under consideration. This is similar to the approach adopted by both MIL-STD-499B and EIA/IS-632.

6.4.5 Additional Material and Requirements

Annex A to IEEE 1220 is an informative summary of the role of systems engineering within an enterprise. This annex does not add additional requirements to the IEEE standard; it merely provides the reader with some background information.

6.4.6 Summary

IEEE 1220 is a detailed and complete systems engineering standard. It is in wide use today and provides the reader with some excellent systems engineering advice and guidance. Unlike EIA/IS-632, IEEE 1220 is not simply a close replica of MIL-STD-499B, but rather a comprehensive coverage of the topic. It provides details of the system life cycle and how the different phases are covered by systems engineering. In addition, it presents a slightly more involved systems engineering process than has been considered in this text and provides the reader with some guidance on the content of the SEMP. IEEE 1220-1998 should be strongly considered when specifying a systems engineering standard.

6.5 ANSI/EIA-632-Processes for Engineering a System

ANSI/EIA-632 was developed jointly by the EIA and the International Council on Systems Engineering (INCOSE) and released in 1999. ANSI/EIA-632 was in development since the release of the interim standard IEA/IS-632 in December 1994. The standard represents a significant change from the interim standard and focuses on the enterprise and its policies and procedures rather than a focusing narrowly on a particular system development being undertaken by that enterprise. This broad focus sets ANSI/EIA-632 apart from the other systems engineering standards currently in use.

ANSI/EIA-632 introduces five broad categories that encapsulate 13 processes required to successfully and completely engineer a system. Within these 13 processes, there are 33 requirements set out in the standard. Not all of the 33 requirements will be relevant or required in every situation;

therefore, the requirement to tailor ANSI/EIA-632 to individual development project remains.

The standard was designed to help enterprises develop policies and procedures relevant to their roles in engineering systems within their industries. Projects being conducted by the enterprise are then responsible for further development of company polices and procedures into project-specific plans, schedules, and so on. By operating at the enterprise level, ANSI/EIA-632 avoids being a prescriptive standard and concentrates on what needs to be done to successfully and completely engineer a system. Figure 6.4 [4] illustrates.

The organization and approach of ANSI/EIA-632 provides a very detailed framework upon which systems engineering enterprises can base their company policies and procedures.

6.5.1 ANSI/EIA-632 Processes

Figure 6.5 [5] shows the 13 processes contained in ANSI/EIA-632 grouped into the five broad categories described in the standard. The categories and processes are arranged into a flow diagram that effectively describes the ANSI/EIA-632 logic. An interesting point to note is that the analysis-synthesis-evaluation loop exists only as a subset of the overall ANSI/EIA-632 logic. This is because ANSI/EIA-632 is an enterprise-wide standard and must address more than the classic systems engineering processes.

A majority of the standard is focused on describing the 13 processes and the 33 requirements that flow out of those processes. Notes and examples are used throughout to clarify the intent of the standard.

ANSI/EIA-632 focuses on process rather than method. The methods for achieving the requirements are expected to come from enterprise policy and procedures. Each of the process areas shown in Figure 6.5 is dealt with separately by considering the inputs and outputs of the process and the requirements within the process. Figure 6.5 shows the major inputs and outputs.

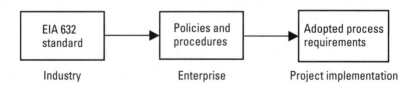

Figure 6.4 Application of ANSI/EIA-632 [4].

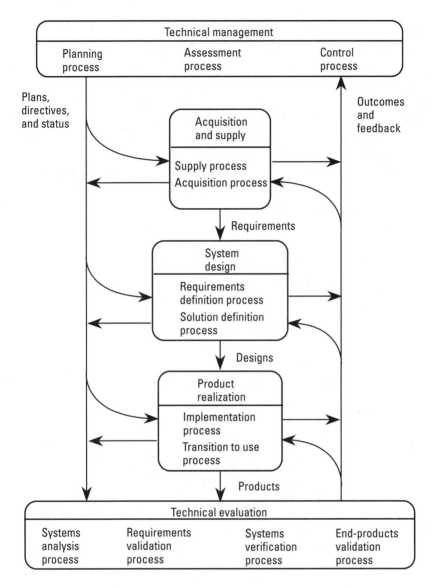

Figure 6.5 ANSI/EIA-632 processes [5].

6.5.2 ANSI/EIA-632 Requirements

Guidance and explanations for each of the 33 requirements is provided in the standard, along with an example wording of the requirement. The reader is

expected to examine each requirement for relevance and tailor the requirements where necessary. Annex C of ANSI/EIA-632 provides the reader with a
process task outcome table that details the representative tasks that will need
to be performed for each of the requirements and the expected outcomes
from those tasks. This annex represents a prescriptive description of the 33
requirements but aids in the understanding of the requirements. Readers are
able to gain an excellent insight into how the requirements might be satisfied
by reading this annex. The following lists the 33 requirements by process
category [6]. For further detail, the reader is referred to the full standard.

- Supply process requirements
    1. Product supply

- Acquisition process requirements
    2. Product acquisition
    3. Supplier performance

- Planning process requirements
    4. Process implementation strategy
    5. Technical effort definition
    6. Schedule and organization
    7. Technical plans
    8. Work directives

- Assessment process requirements
    9. Progress against plans and schedules
    10. Progress against requirements
    11. Technical reviews

- Control process requirements
    12. Outcomes management
    13. Information dissemination

- Requirements definition process requirements
    14. Acquirer requirements
    15. Other stakeholder requirements
    16. System technical requirements

- Solution definition process requirements
    17. Logical solution representations

18. Physical solution representations
19. Specified requirements

- Implementation process requirements
  20. Implementation

- Transition to use process requirements
  21. Transition to use

- Systems analysis process requirements
  22. Effectiveness analysis
  23. Trade-off analysis
  24. Risk analysis

- Requirements validation process requirements
  25. Requirement statements validation
  26. Acquirer requirements validation
  27. Other stakeholder requirements validation
  28. System technical requirements validation
  29. Logical solution representations validation

- System verification process requirements
  30. Design solution verification
  31. End product verification
  32. Enabling product readiness

- End products validation process requirements
  33. End products validation

6.5.3 ANSI/EIA-632 Concepts

ANSI/EIA-632 defines a system as comprising of hardware, software, people, facilities, data, materials, services, and techniques. In addition to this definition, ANSI/EIA-632 proposes that a system consists of the end product that is the classic deliverable, but also the so-called enabling products that allow the end products to be produced. The categories of enabling products and some examples provided in ANSI/EIA-632 are listed in Table 6.2.

The other major concept introduced in ANSI/EIA-632 is that of the building block. This shows that each end product (of which there may be many per system) can be broken into subsystems. Each subsystem can be considered a system in its own right and dealt with in accordance with

Table 6.2
ANSI/EIA-632 Enabling Products and Examples [7]

Enabling Products	Examples
Development	Plans, schedules, policies, procedures, tools, models, interfaces, and development personnel
Production	Plans, schedules, policies, procedures, facilities, tools, equipment, materials, and production personnel
Test	Plans, schedules, policies, procedures, mockups, test equipment, test facilities, simulations, models, and test personnel
Deployment	Plans, schedules, policies, procedures, packaging, storage, handling, installation, transportation, sites, and installation personnel
Training	Plans, schedules, policies, procedures, simulators, training aids, courses, facilities, and trainers
Support	Plans, schedules, policies, procedures, tools, repair equipment, support facilities, maintenance manuals, diagnostic equipment, and repair personnel
Disposal	Plans, schedules, policies, procedures, disposal sites, special equipment, and disposal personnel

ANSI/EIA-632. This forms the basis of the top-down development concept supported by ANSI/EIA-632, which breaks the end products into increasing levels of details from layer 1 (system level) down to as many layers as required by the development process. Once the system has been completely broken down in this way, design decisions can be made regarding how each subsystem is designed. Figure 6.6 [8] shows the building-block concept introduced in ANSI/EIA-632.

ANSI/EIA-632 then goes on to describe the bottom-up realization concept, which details how the very low-level layers are progressively integrated and tested until the system-level end product is produced. The top-down development and bottom-up realization concepts are not new and form the basis for most popular systems engineering thinking on system development (as described in Chapters 1 and 4).

Table 6.3 shows the enterprise-based life cycle as detailed in Annex B to ANSI/EIA-632 and compares it with the Blanchard life cycle used in this text. The engineering life cycle detailed in ANSI/EIA-632 does not include the utilization phase activities.

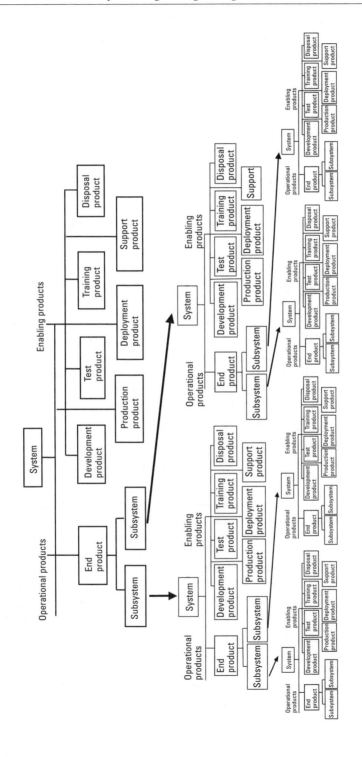

Figure 6.6 ANSI/EIA-632 building-block concept [8].

Table 6.3
ANSI/EIA-632 Life-Cycle Activities Compared with Those Used in This Text

ANSI/EIA-632 Life-Cycle Activities	Life-Cycle Activities
Assessment of opportunities investment decision	Need
System concept development	Conceptual design
Subsystem design and predeployment	Preliminary design, detailed design, and development, construction, and production
Deployment, operations, support, and disposal	Operational use and system support

6.5.4 ANSI/EIA-632 Annexes

ANSI/EIA-632 provides a number of useful annexes. Annex A is a standard inclusion listing the glossary of terms used throughout the document. Annex B provides additional detail on the engineering life cycle used throughout the standard. Annex C provides a very useful description of each of the requirements in terms of representative tasks necessary to achieve the requirements and the expected outcomes from each of the representative tasks.

Annex D provides a description of the documentation that may be required from a system development project. The documentation is broken into the following categories:

- *Source documents.* These documents include concept documentation, such as maintenance concepts, operational concepts, and disposal concepts.

- *Technical documents.* This section provides a comprehensive list of technical documentation that could be expected to flow from the system development effort. These documents are listed alphabetically and include the systems engineering management documents and the systems engineering process documents described in this book.

- *Enterprise or project documents.* These documents provide guidance and constraints for the development effort. These documents typically include enterprise policies and procedures.

Annex E provides details of the technical reviews and audits that could be expected throughout the system development effort. This detail is provided relative to the engineering life-cycle activities. The description of the reviews and audits provides details of what each review and audit should aim to achieve.

Annex F describes the activities needed to fully define the end products. The activities depend on whether or not the end products are already known. The end products will be known if COTS equipment is to be used. However, end products that are to be designed require a slightly different approach to definition.

Annex G provides an explanation of the relationship between the many sources of requirements and the eventual solution. The sources of requirements include acquirer requirements, technical requirements, and derived technical requirements. These are then assigned to logical and physical representations that eventually lead to a set of specified requirements.

6.5.5 Summary

ANSI/EIA-632 is an excellent systems engineering document that has progressed significantly since the release of its predecessor. The scope of ANSI/EIA-632 has grown to incorporate the entire enterprise rather than just the system development project. The standard is based on five process categories that include 13 individual processes. ANSI/EIA-632 breaks down these 13 processes into 33 requirements that must be performed to develop a product. These requirements are designed to be incorporated into enterprise policies and procedures. Depending on the specific nature of the system development effort, some of the 33 requirements may be tailored out of the standard.

ANSI/EIA-632 is the latest systems engineering standard and provides a very solid framework upon which systems engineers can base system development efforts. As the most recent systems engineering standard available, practitioners should strongly consider referring to, and requiring, ANSI/EIA-632 during system development projects, especially if their industry has adopted the standard and their enterprise has progressed sufficiently to provide the necessary process and procedure to support the use of the standard.

6.6 Other Useful Documents

The following section lists and describes some additional sources of specific information relating to systems engineering topics. Most of the documents listed are available on the Internet, following a simple search using their document number.

6.6.1 Technical Reviews and Audits

MIL-STD-1521B [9] is an old standard that has now been cancelled by the U.S. DoD. Its content has been summarized and divided between MIL-STD-499B and MIL-STD-973. Despite this, it remains an excellent reference source for technical reviews and audits. It provides a thorough coverage of a comprehensive range of reviews and audits, and links the reviews and audits to the systems engineering life cycle presented in the systems engineering standard of the day.

MIL-STD-1521B provides useful descriptions of the types of issues to be covered at each of the reviews and the criteria that should be met prior to entering and exiting each review. In addition, it provides checklists and templates that may be of assistance during the technical review and audit program.

6.6.2 Systems Engineering Standards

ECSS-E-10A [10] is part of a set of standards issued by the European Cooperation for Space Standardization and tailored for use in space applications. It provides a useful coverage of the requirements engineering process, configuration management, systems engineering products, and interfaces with other disciplines. The standard is written as a set of statements of what must be done in systems engineering rather than describing how to do it.

6.6.3 Configuration Management

ANSI/EIA-649-1998 [11], MIL-STD-973 [12], and MIL-HDBK-61A(SE) [13] are three standards that describe configuration management practices and processes. They provide a great deal of detail on the process described in Chapters 2 to 5. MIL-STD-973 is regarded as the classic configuration management standard and remains an excellent source of reference, although it is becoming a little dated in some regards. The more recent configuration management standards cover current configuration management challenges, such as configuration management of electronic documentation and media.

6.6.4   Specification Standards

MIL-STD-490A [14] has been cancelled and replaced by MIL-STD-961D [15]. However, as is often the case, MIL-STD-490A remains an excellent source of information on specifications. The standards introduce a hierarchical set of specifications from the A-level system specification to the E-level materials specification. The relationship of these standards to functional (A-level specification), allocated (B-level specifications), and product baselines (C-, D-, and E-level specifications) are also explained. The standard also describes the many different types of development and product specifications, including prime item and critical item development and product specifications. The document provides content information for each of the specifications in an appendix.

MIL-STD-961D has superseded MIL-STD-490A and is written to describe the content and requirements of generic specifications rather than content of specific specifications as outlined in its predecessor. With that in mind, reference to both of the specifications can assist in an understanding of the types and content of technical specifications required on given system developments.

6.6.5   Work Breakdown Structures

Military handbook MIL-HDBK-881 [16] is an excellent source of information on work breakdown structures and represents a conversion of the earlier military standard MIL-STD-881B, *Work Breakdown Structures for Defense Materiel Items.* MIL-HDBK-881 effectively describes the relationship between the program WBS and the contract WBS. It also contains detailed information on the purpose, structure, and content of a range of example programs, including an aircraft system that proved useful in this text.

6.7   Capability Maturity Models

CMMs are becoming prevalent across a number of disciplines, including systems engineering and software engineering in particular. Systems engineering capability maturity models are designed to compliment systems engineering standards, such as those reviewed in the previous section, and as such form an important systems engineering tool. While engineering standards such as ANSI/EIA-632 and IEEE 1220 explain what to do with respect to systems engineering processes, CMMs aim to provide a basis for determining how well the processes are defined and implemented [17].

In general, CMMs should be used by organizations involved in systems engineering to evaluate the systems engineering processes currently in use and provide an insight into improving those processes. Additionally, systems engineering CMMs also allow organizations to design and develop new processes.

There are currently three systems engineering CMMs in existence, although these models are being rationalized and combined into a single product. This single product is also in the process of being incorporated into a broader CMM called the CMM integration (CMMI), covering a range of disciplines including systems engineering. With three systems engineering CMMs in existence, confusion regarding their respective sources, roles, and differences prevails. Figure 6.7 shows the three extant systems engineering CMMs, indicates the systems engineering standards that have influenced the development of each CMM, and shows how the CMMs are currently being combined to play a role in the CMMI.

The first systems engineering CMM was produced at the Carnegie Mellon University's Software Engineering Institute (SEI) and is called the systems engineering capability maturity model (SE-CMM). Version 1.1 is the current version of the SE-CMM, released in 1995. The SE-CMM

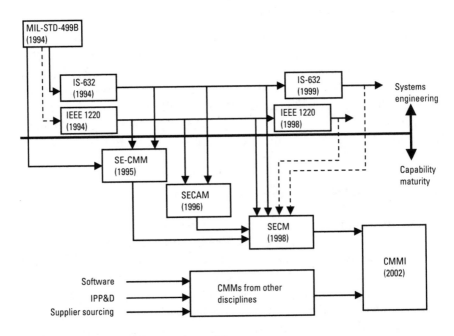

Figure 6.7 Extant capability maturity models.

was written to be consistent with MIL-STD-499B, EIA/IS-632, and IEEE1220-1994.

Almost in parallel with the SEI effort, the INCOSE was authoring a similar product called the systems engineering capability assessment model (SECAM). It too was written within the bounds of EIA/IS-632 and IEEE1220-1994. The current version of the SECAM is version 1.5, which was released in 1996.

The EIA has more recently released an interim standard called the systems engineering capability model (SECM) or EIA/IS-731. This is an amalgamation of the SE-CMM and the SECAM. It was released in 1998 and is cognizant of the CMMs and systems engineering standards in existence at that time. Its release date of May 1998 predates both current systems engineering standards (IEEE1220-1998 released in December 1998 and EIA/ANSI-632 released in January 1999) by a matter of months. The SECM does, however, acknowledge the existence of both IEEE1220-1998 and EIA/ANSI-632 and makes a commitment to maintain consistency with both standards (that commitment is indicated by dotted lines in Figure 6.7). With the release of the SECM, the SE-CMM and SECAM effectively became superseded. The SECM has been selected to form the systems engineering component of the CMMI. As the CMMI becomes more widely accepted, the SECM will cease to be supported by EIA.

6.8   SEI—Systems Engineering Capability Maturity Model

The aim of this section is not to comprehensively detail the content of the SEI SE-CMM but rather to discuss its existence and intent. By doing so, the intent and application of CMMs in general is introduced.

The SE-CMM describes the elements of an organization's processes that must exist to ensure a requisite level of systems engineering capability within that organization. The model was developed to allow organizations to perform a self assessment of their current ability to effectively translate a customer's requirements into a product that either meets or exceeds those requirements. Based on this self assessment, the organization should be able to improve their systems engineering capability.

Although the SE-CMM discourages use of the model as a tool for supplier selection, the systems engineering capability of potential contractors would certainly be of interest to the customer during the selection process. The SE-CMM certainly focused on process improvement via self assessment.

6.8.1   SE-CMM Foundation

The SE-CMM believes that people, process, and technology are the main components of organizational capability, which, in turn, determines product and service quality. This foundation is shown in Figure 6.8 [18].

The SE-CMM concentrates on the process aspect of this capability, as process tends to integrate people and technology. Focusing on process also helps to improve performance prediction and actual performance. The model is aimed at being a guide to organizations rather than a prescription of how to improve systems engineering capability. Although the SE-CMM is a stand-alone document, the model is also cognizant of MIL-STD-499B, EIA/IS-632, and IEEE 1220-1994.

6.8.2   Process Areas

The SE-CMM uses process areas (PAs) to describe the major areas essential to sound systems engineering. There are 18 PAs, divided into three groups called process categories: engineering, project, and organization. The 18 PAs divided into the process categories are summarized in Table 6.4.

A *base practice* is an engineering or management activity that is performed to address the requirements of a PA. The SE-CMM comprehensively details the 18 PAs and describes the mandatory base practices considered essential for the achievement of the given PA. By studying the PAs and their respective base practices, organizations can initially assess and then improve their capabilities in that area.

6.8.3   Capability Levels

To make a self assessment, an organization will investigate its ability to perform each of the PAs and record its performance using the SE-CMM

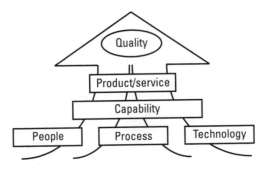

Figure 6.8   SE-CMM foundation.

Table 6.4
SEI SE-CMM Process Areas

Engineering Process Area	Project Process Area	Organizational Process Area
Analyze candidate solutions	Ensure quality	Coordinate with suppliers
Derive and allocate requirements	Manage configurations	Define organization's systems engineering process
Evolve system architecture	Manage risk	Improve organization's systems engineering process
Integrate disciplines	Monitor and control technical effort	Manage product line evolution
Integrate system	Plan technical effort	Manage systems engineering support environment
Understand customer needs and expectations		Provide ongoing knowledge and skills

capability levels. Based on the performance, the organization can work on ways to improve its capability. Six generic levels of capability are described [19]:

- *Level 0* (not performed) basically indicates that some of the base practices associated with PAs are being completed, but the performance is inconsistent.

- *Level 1* (performed informally) indicates that all of the base practices associated with a PA are being performed somewhere within the organization, but planning or tracking are not being performed. The quality of the performance is often based on the skills and knowledge of individuals rather than the organization. In this case, the performance of the base practices will often not be repeatable.

- *Level 2* (planned and tracked) demonstrates that the performance of the base practices is planned and tracked by the organization. Poor performance is noted and is subject to corrective action. The planning and tracking helps to ensure that the performance is repeatable.

- *Level 3* (well defined) indicates that base practices are performed throughout the organization in accordance with approved and documented processes. Performance is continually measured and monitored (to determine whether any adjustments are required). Planning and management of daily activities are performed and the organization is committed to long-term process improvement.

- *Level 4* (quantitatively controlled) should be assigned where measurable goals are established and the performance of base practices is analyzed against these goals. This level of capability enhances an organization's ability to predict performance.

- *Level 5* (continuously improving) is assigned when qualitative and quantitative goals are established based on long-term strategic direction. The performance of the processes is continually monitored and improved with the organizational goals in mind. Innovation and new technology are often part of the performance at this level of capability.

6.8.4 Summary

The SE-CMM provides a model for organizations to use in determining their current levels of maturity in terms of systems engineering capabilities and improving this maturity, if possible. The model is based on process areas and the performance of base practices to varying levels of capability.

6.9 CMM Integration

There are currently three CMMI [20] products available covering combinations of the following four disciplines:

1. Systems engineering (SE);
2. Software engineering (SW);
3. Integrated product and process development (IPPD);
4. Supplier sourcing (SS).

The three CMMI products are offered with the following combination of disciplines:

1. SE/SW;
2. SE/SW/IPPD;
3. SE/SW/IPPD/SS.

Organizations wishing to adopt the CMMI must determine which discipline areas are relevant to them and look towards one of the three CMMI products to cover their requirements. Each product is offered in a staged or continuous representation. By design, the granular information contained in the two representations is virtually identical. However, each of the representations provides benefits that will be valued differently by different organizations.

In the continuous representation of a CMMI model, the summary components are process areas. Specific goals are implemented by specific practices within each process area. Also contained in the continuous representation of a CMMI model are generic goals that are implemented by generic practices. Specific goals and practices are unique to individual process areas, whereas generic goals and practices apply to multiple process areas. Each practice belongs to only one capability level. For example, to satisfy capability level 2, an organization must satisfy the specific goals and level 2 practices for that process area, as well as the level 2 goals for that same process area.

In the staged representation, the summary components are maturity levels. Within each maturity level, process areas contain goals, common features, and practices, which serve as guides to what to implement to achieve the goals of the process area. Practices are categorized into common features:

- *Commitment to perform*—practices that ensure that the process is established and will endure. These practices typically involve establishing organizational policies and procedures.

- *Ability to perform*—practices that establish the necessary conditions for implementing the process completely. These practices typically involve plans, resources, organizational structures, and training.

- *Activities performed*—practices that directly implement a process. These practices distinguish a process area from others.

- *Directly implementing*—practices that monitor and control the performance of the process. These practices typically involve designating work products of the process under configuration management, monitoring and controlling the performance of the process against the plan, and taking corrective action.

- *Verifying implementation*—includes practices that ensure compliance with the requirements of the process area. These practices typically involve reviews and audits.

Endnotes

[1]   MIL-STD-499B, *Military Standard—Systems Engineering—Draft*, Washington, D.C.: U.S. Department of Defense, 1994.

[2]   MIL-STD-499B, *Military Standard—Systems Engineering—Draft*, Washington, D.C.: U.S. Department of Defense, 1994.

[3]   IEEE-STD-1220-1998, *IEEE Standard for Application and Management of the Systems Engineering Process*, New York: IEEE Computer Society, 1998.

[4]   ANSI/EIA-632-1998, *EIA Standard—Processes for Engineering a System*, Arlington, VA: Electronic Industries Association, 1999.

[5]   ANSI/EIA-632-1998, *EIA Standard—Processes for Engineering a System*, Arlington, VA: Electronic Industries Association, 1999.

[6]   ANSI/EIA-632-1998, *EIA Standard—Processes for Engineering a System*, Arlington, VA: Electronic Industries Association, 1999.

[7]   ANSI/EIA-632-1998, *EIA Standard—Processes for Engineering a System*, Arlington, VA: Electronic Industries Association, 1999.

[8]   ANSI/EIA-632-1998, *EIA Standard—Processes for Engineering a System*, Arlington, VA: Electronic Industries Association, 1999.

[9]   MIL-STD-1521B, *Military Standard—Technical Reviews and Audits for Systems, Equipments, and Computer Software*, Washington, D.C.: U.S. Department of Defense, 1985.

[10] ECSS-E-10A, *Space Engineering-System Engineering*, Noordwijk, the Netherlands: European Cooperation for Space Standardization, 1996.

[11] ANSI/EIA-649-1998, *EIA Standard—National Consensus Standard for Configuration Management*, Arlington, VA: Electronic Industries Association, 1998.

[12] MIL-STD-973, *Military Standard—Configuration Management*, Washington, D.C.: U.S. Department of Defense, 1992.

[13] MIL-HDBK-61A(SE), *Military Handbook—Configuration Management Guidance*, Washington, D.C.: U.S. Department of Defense, 2001.

[14] MIL-STD-490A, *Military Standard—Specification Practices*, Washington, D.C.: U.S. Department of Defense, 1985.

[15] MIL-STD-961D, *Department of Defense Standard Practice for Defense Specifications*, Washington, D.C.: U.S. Department of Defense, 1995.

[16] MIL-HDBK-881, *Department of Defense Handbook—Work Breakdown Structure*, Washington, D.C.: U.S. Department of Defense, 1998.

[17] For more information on the CMM, see the following:

EIA/IS 731.1, *Systems Engineering Capability Model*, Arlington, VA: Electronic Industries Association, May 1998.

Carnegie Mellon University Software Engineering Institute, *The Capability Maturity Model: Guidelines for Improving the Software Process*, Reading, MA: Addison-Wesley, 1999.

[18]   SECMM-95-01, *Systems Engineering Capability Maturity Model*, Version 1.1, Carnegie Mellon University, Pittsburgh, PA: Software Engineering Institute, 1995, p. 2-2.

[19]   SECMM-95-01, *Systems Engineering Capability Maturity Model*, Version 1.1, Carnegie Mellon University, Pittsburgh, PA: Software Engineering Institute, 1995, p. 2-28.

[20]   Further information on CMMI can be found in the following:

CMMI-SE/SW/IPPD/SS, V1.1, *Capability Maturity Model Integration (CMMI) (Version 1.1) for Systems Engineering, Software Engineering, Integrated Product and Process Development, and Supplier Sourcing*, Carnegie Mellon Software Engineering Institute; Pittsburgh, PA., March 2002.

CMMI-SE/SW/IPPD, V1.1, *Capability Maturity Model Integration (CMMI) (Version 1.1) for Systems Engineering, Software Engineering, and Integrated Product and Process Development*, Carnegie Mellon Software Engineering Institute; Pittsburgh, PA, December 2001.

CMMI-SE/SW, V1.1, *Capability Maturity Model Integration (CMMI) (Version 1.1) for Systems Engineering and Software Engineering*, Carnegie Mellon Software Engineering Institute; Pittsburgh, PA, December 2001.

7

Systems Engineering Process Tools

Systems engineering tools are available to assist both the systems engineering management function and the systems engineering processes, as well as assisting related disciplines such as project management. Chapter 6 introduced a number of tools that may assist with systems engineering management and processes. This chapter examines tools to assist in the conduct of systems engineering processes. Some analysis tools have been described in Chapters 2, 3 and 4—this chapter focuses on requirements engineering. Schematics, modeling, and simulation are detailed as tools that can assist the synthesis aspect of the systems engineering processes. Trade-off analysis is also detailed as a potential tool for use during the evaluation stage of the process.

7.1 Analysis Tools—Requirements Engineering

Some analysis tools have been described in Chapters 2, 3, and 4, including the context diagram, FFBD, RBS, and N2 diagrams. Other tools include structured analysis, the data flow diagram, control flow diagram, IDEF diagram, behavior diagram, action diagram, state/mode diagram, process flow diagram, function hierarchy diagram, state transition diagram (STD), entity relationship diagram, structured analysis and design, object-oriented analysis, unified modeling language, structured systems analysis and design (SSADM), and quality function deployment. Each of these techniques focuses on gathering requirements in a formal systematic way, which is normally referred to generically as *requirements engineering* [1]. We focus on the philosophy of requirements engineering here because it is such an essential element of system development, and we leave the reader to consult the many excellent references for the more detailed tools [2].

We have discussed many general aspects of requirements engineering in our consideration of conceptual design and preliminary design (Chapters 2 and 3). It is such an essential tool of systems engineering, however, that we describe it in much more detail in this chapter.

7.1.1   What Is a Requirement?

A requirement is a statement of a system service or a constraint placed on the system. At the system level, the requirements document (system specification) might therefore describe the following:

- The services and functions that the system should provide;
- The constraints under which the system should operate;
- Overall properties of the system;
- Definitions of other systems that the system must integrate with and interface to;
- Information about the application domain of the system (for example, how to carry out particular types of computation);
- Constraints on the way in which the system is to be developed.

In general, a requirement should be a statement of what a system should do, rather than how it should do it. This is often too simplistic in practice, however, because it may be essential that the system should perform a function in a particular way to ensure interoperability or to meet extant standards. Additionally, a specific statement is often less easily confused than an abstract statement of the problem, which can lead to confusion in the minds of lower level designers. Finally, the systems engineers conducting conceptual design of the system are often the domain experts and are therefore best placed to state how the system should be developed and how it should operate.

7.1.2   Requirements Engineering

There is no standard requirements engineering process—in fact, few organizations have mature, explicitly defined, standardized processes for requirements development. The requirements engineering process is a set of activities intended to derive, validate, and maintain a set of system requirements. Activities include requirements elicitation and generation, requirements analysis and negotiation, requirements allocation and maintenance/management, and requirements validation. Figure 7.1 illustrates the iterative manner in which these activities are conducted.

The complete requirements engineering process should be described in the SEMP and, in addition to the detail described in this section, should include what activities are to be carried out, who is responsible for them, the inputs and outputs, and what tools are to be used.

7.1.2.1 Requirements Elicitation and Generation

Requirements elicitation and generation activities involve working with customers and end users to find out about the problem to be solved, the required performance of the system, constraints, and so on. Elicitation generally involves assessment of the range of system profiles from manufacturing to disposal within the range of appropriate environmental conditions, as well as identification of constraints and suitable regulatory and statutory

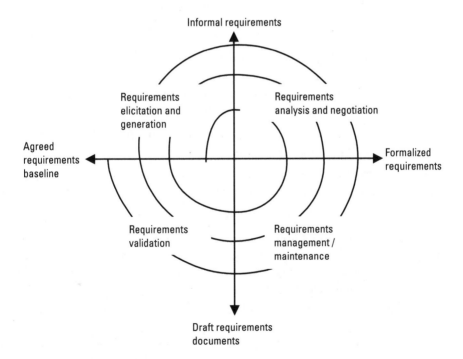

Figure 7.1 Iterative requirements engineering activities.

requirements. Unfortunately, elicitation is not always a simple knowledge-transfer process. In reality, users rarely have a clear view of their requirements, different people in an organization have conflicting requirements, users are rarely aware of the limitations that technology will have on their expectations, and so on.

There are four dimensions to requirements elicitation:

1. *Understanding the application domain.* Requirement writers must understand the general application domain of the system to be implemented.

2. *Understanding the specific problem.* The specific customer problem must be understood—the general application domain understanding must be specialized and extended.

3. *Understanding the business.* Requirement developers must understand how the system to be developed contributes to the business or organization.

4. *Understanding the needs and constraints of system stakeholders.* These specific needs must be understood, as there may be several different criteria by which the final system is judged.

Requirement engineers face a number of problems:

- Application-domain information is not always collected in one place and may involve specialist terminology.

- People who understand the problem to be solved are often too busy to help specify the new system.

- Organizational issues and political factors may influence system requirements to satisfy personal agendas.

Stakeholders do not always help—they sometimes do not really know what they want, they often have no desire to compromise their unrealistic demands when faced with cost, and they have different requirements that are often expressed in different ways.

Requirements can be identified by the following:

- *Interviews, surveys, and questionnaires.* These techniques offer a formal process through which requirements may be garnered from individual users. Their use should be considered carefully, however, in that respondents only provide useful information in response to focused, well-balanced questions, which implies that a measure of requirements engineering has been conducted before the interviews are held or before the surveys and questionnaires are sent out. All three processes can also be very time consuming, which again requires that careful thought is given to the questions to ensure that responses add to an understanding of the systems requirements, rather than add additional noise.

- *Structured workshops, brainstorming, and problem-solving sessions.* A popular way of avoiding workplace distractions is to hold structured user-requirements workshops, usually in a convenient off-site location where users can devote their attention to articulating their needs. Again, however, best results are achieved if workshop attendees have some basis for discussion—that is, a first iteration of requirements engineering should be conducted before the workshop is held. Conducting a workshop that is not informed by some well-focused preliminary documents normally leaves users with a poor opinion of the new system even before conceptual design is well under way. Another way to ensure that users make valuable contributions is to conduct a period of briefing/training before beginning the requirements elicitation.

- *Observation of users in the workplace.* Time and motion studies may not always be in favor and may not always be a suitable way of extracting requirements, but it is recommended that systems engineers use the context of the workplace to develop a comprehensive understanding of user requirements. Observing users in their natural environment often identifies requirements that are not stated by the user in a more formal workshop setting. Users are not always consciously aware of all of their work practices that are relevant to the new or modified system.

- *Market analysis.* Many products do not have a readily available, homogeneous set of users, such as those that we have been able to identify within ACME Air for our new aircraft. Consumer products such as motor vehicles require the set of users to be more closely defined by identifying the requirements of a wide range of users. Because these requirements will cover a broad range and will often conflict, market analysis techniques may assist in separating out a subset of achievable requirements that maximize the potential market for the product.

- *Documentation review.* Of both business and technical documents—few organizations have info-rich documents, although many organizations have standard operating procedures that may describe in some detail how particular operations are conducted

- *Prototypes, simulations, and use cases.* One of the difficulties with asking users about their requirements for the new system is that users will probably have been trained on an existing system and may have difficulty in setting extant operating procedures aside when considering the new system requirements. Prototypes, simulations, and use cases (sometimes generically called *storyboarding*) assist in providing operators with a new frame of reference, particularly at the beginning of the requirements-elicitation process where they may have difficulty in breaking from existing practices. Requirements validation is also greatly enhanced by the ability of requirements engineers to mock up a system that meets the collected requirements and then present that to users for evaluation. Prototypes and simulations can be very complex or they can be as simple as a series of role-playing exercises in which users are asked to describe how they would imagine the system operating under particular circumstances. Use cases could be considered to be a form of paper-based prototyping in which stakeholders describe their requirements in a range of typical operations for the system.

- *Competitive system assessment and trials.* Like prototypes and simulations, competitive system assessment and trails are aimed at providing users with some tangible manifestation of early requirements so as to assist in requirements elicitation and validation. One simple, quick way of providing depth to early requirements is to identify existing systems that meet a subset of early user requirements and ask users to evaluate them in the context of their need, goals, and objectives. Users should be asked to describe as many use cases as required to cover every anticipated role of the new system—more detail is invariably better than less.

However it is gathered, each requirement is to be examined as it is collected or derived to ensure that it has a number of important properties. It should be all of the following:

- Unique in that it does not have the same purpose as any other requirement or a combination of several requirements;

- Complete;
- Correct;
- Unambiguous;
- Achievable using extant technologies and manufacturing procedures;
- Independent of the method of implementation;
- Traceable with respect to at least one higher level requirement;
- Verifiable in that it can be proved that the system meets or possesses the requirement.

The statement of each requirement is not sufficient on its own, however. To support analysis and management, requirements should be recorded with associated information, including the following:

- *Unique identifier.* Each requirement must be able to be identified as a unique statement. When a requirement is recorded, it must be allocated a unique identifier that remains with the requirement throughout the design process. This is essential for requirements management and for requirements traceability in particular. While managing the allocation of unique numbers is difficult in a paper-based system, the process is straightforward when conducted within the database engine underlying an automated requirements-management tool. Each functional requirement also has associated with it a performance and verification requirement. While these require separate identifiers, it is useful to use a system whereby the association between the three types of requirements is immediately obvious. One recommended system is shown in Figure 7.2, where the same stem number is employed for all three types and the letters P and V are appended to denote the performance and verification requirements associated with the functional statement.

- *Priority.* This value is allocated from a user perspective and indicates how important the achievement of each requirement is in the context of the overall system. Users are tempted to state all requirements as having the highest priority, but they should be encouraged to provide some flexibility within which design can be conducted. The most important (mandatory) requirements are often stated as *shall;* the next most important are stated as *shall, where practicable;* the next most important are stated as *preferred* or *should;* and the next most important are stated as *may.* These levels are often simplified to

ID	Requirement Statement	Interface	Priority	Risk	Criticality	Feasibility	Originator	Type
12.1.2	Functional statement							
12.1.2P	Performance statement							
12.1.2V	Verification statement							
12.1.3	Functional statement							
12.1.3P	Performance statement							
12.1.3V	Verification statement							
12.1.4	Functional statement							

Figure 7.2 Suggested format for a paper-based specification.

a three-level (mandatory, highly desirable, desirable) or a two-level (mandatory, optional) scheme. When initially collecting requirements, it may be sufficient to ask users to rate the requirement's priority as high, medium, or low. A more detailed system could ask them to rate priority on a scale of 1 to 10, which would allow some numerical analysis of priority during subsequent design, if that is desirable. In baseline specifications, priority can therefore be allocated to the requirement in two ways—either specified in the text of the statement using shall, should, and may, or all statements could be stated using terms such as shall and an indication of priority could be attached on a numerical scale or in the form of high, medium, or low.

- *Criticality.* It is also useful to ask the user to identify those requirements that are critical in terms of their contribution to system performance. While seemingly related to priority, this category identifies show stoppers—those requirements that are essential to the success of the overall system. In other words, a requirement that is deemed critical must be met or possessed by the system or else the system cannot be introduced into service. In some cases it may be useful to have a rating system, but it is most often sufficient to identify each requirement in a binary fashion as critical or noncritical.

- *Feasibility.* Most system developments and acquisitions have some elements that are to be based on new technology or require the development of new manufacturing and production techniques. It may be useful in these cases to establish the feasibility of meeting the requirement based on an understanding of current technology and procedures. The major use for this category is in the functional analysis tasks associated with the development of the draft system specification—it is most likely that the utility of the category beyond that stage is marginal, and it may well not be included in the approved system specification.

- *Risk.* Each requirement contributes to overall system risk in its own way, and the management of risk is a critical component of systems engineering and project management. We discuss risk management in some detail in Section 5.4.

- *Source.* There are likely to be a number of sources for requirements. The originator (person, organization, document, or process) must be recorded to ensure that all requirements can be justified. Additionally, sources must be recorded so that they can be consulted when compromises are required during functional analysis.

- *Type.* While not essential, it is very useful to attach to each requirement a tag associated with the type of requirement. Example are such types as input, output, external interfaces, reliability, availability, maintainability, accessibility, environmental conditions, ergonomics, safety, security, facility, transportability, training, documentation, testing, quality provision, policy and regulatory, compatibility with existing systems, standards and technical policies, conversion, growth capacity, and installation. The type field is most useful because it allows the requirements database to be viewed by a large number of designers and stakeholders for a wide range of uses. For example, maintainers could review the database by asking to be shown all the maintenance requirements, engineers developing test plans could extract all T&E requirements, and so on.

- *Rationale.* As described in Section 2.4.5, each requirement must have associated with it the reason for its existence.

- *History.* This portion of the database records when the requirement was raised, by what means it was raised, and by whom. It also identifies any changes that are made to the requirement throughout the design process and records when the requirement was modified and by whom.

- *Relationship to other requirements.* The relationship with other requirements must be identified to assist in forward and backward traceability.

- *Other information.* If desired, each requirement should be stored with reference to the date of raising the requirement, the status (proposed, reviewed, accepted, rejected, and so on), and comments.

7.1.2.2 Requirements Analysis and Negotiation

Not only is it difficult to extract all of the relevant requirements from users, but those that can be extracted tend to be relatively unstructured and random. In addition, they are often incoherent and redundant in that a single requirement may be stated in a number of slightly different ways depending on the viewpoints of various users. Requirements collected from users must be therefore be analyzed to obtain an adequate understanding of the requirements and negotiated to agree upon a well-structured, coherent requirements baseline.

Requirements analysis involves identifying incorrect assumptions, ensuring consistency, identify misunderstandings, and so on. Requirements analysis involves a number of activities:

- Requirements are checked for necessity (each requirement must contribute to meeting the system need), completeness, accuracy, compatibility, consistency, traceability, verifiability (each requirement must be able to be tested), appropriate level (mandatory versus optional), and feasibility.

- A set of derived requirements is developed. Derived requirements are not part of the original specification of the system, but are derived in order to design the subsystems. For example, if there is a requirement for a certain degree of reliability, system engineers must derive the requirements for the reliability of each subsystem. Such derived requirements are then allocated to the subsystems.

- System requirements must be allocated to the lower-level subsystems and components.

- Requirements are placed in a structured context using a tool such as an RBS.

7.1.2.3 Requirements Management or Maintenance

Requirements management, or requirements maintenance, is the process by which changes to requirements are managed throughout the system life cycle. Requirements require management for a number of reasons. The large number of detailed requirements in lower level specifications has been derived from a relatively simple statement of the user need; these lower level requirements must be managed as they are developed to ensure that each of them can be justified. User requirements also change over the life of the system due to changes in the business environment as well as the laws and regulations that govern the various aspects of the system. It is often the case that more than 50% of a system's requirements will be modified before it is put into service.

Requirements management involves establishing and complying with a formal procedure for the collection, verification, and traceability of requirements. Additionally, a formal procedure is required for the controlling of changes to requirements.

7.1.2.4 Requirements Validation

Requirements must be validated to be complete, consistent, and unambiguous to certify that the requirements represent an acceptable description of the system that is to be implemented. This process is intended to detect any problems before it is used as a basis for system development. In short, requirements validation is mostly concerned with the question, have we got the right requirements?

The main problem with requirements validation is that there is no existing document that can be used as the basis of the validation. In later phases of the life cycle, a design or a program can be validated against the specification. However, there is no way to demonstrate that the requirement specification is correct. Specification validation therefore really means ensuring that the requirements document represents a clear description of the system for design and implementation and is a final check that the specification meets the needs of stakeholders. Validation is normally conducted by a formal requirements review or a series of reviews.

7.1.3 Requirements Documentation

Writing well-formed requirements can be achieved by using the following guidelines [3]:

- Ensure that each requirement is a necessary, short, definitive statement of functionality.

- Define the appropriate conditions (qualitative or quantitative measures) for each requirement.

- Ensure that each functional requirement is verifiable through the discipline associated with writing a verification statement for every functional statement.

- Avoid overspecification, unnecessary constraints, and unbounded statements.

- Ensure readability by defining standard templates for describing requirements; using natural language simply, consistently, and concisely; using short sentences and paragraphs, as well as lists and tables; supplementing natural language with other descriptions where appropriate, such as equations if they are the most unambiguous way of describing requirements; and using diagrams where appropriate to show relationships between entities.

Requirements must be placed into a structured format. For example, a suggested paper-based format for a specification is illustrated in Figure 7.2, where columns are included to contain most of the information required to be recorded along with the requirements (see Section 7.1.2.1).

Note that it is impossible to include all of the information on a single sheet of paper in a manageable way. In addition to the fields in Figure 7.2, the history and rationale for each functional requirement should be recorded, which results in the table becoming much larger than a single page. The difficulties with paper-based formats for specifications can be overcome with the use of requirements management tools, which are discussed in some detail in Section 7.1.4.

7.1.4   Automated Requirements-Management Tools

Requirements-management tools have been developed to assist in the difficult task of managing the large volume of data associated with the management of requirements. The tools generally provide support in importing and exporting requirements to and from other databases, spreadsheets, and text files; storing and retrieving requirements; visualizing the links between requirements; providing tools that facilitate change control; linking for forward and backward traceability; providing tools for navigating through the database; extracting subsets of requirements; and so on. Some tools also allow for dynamic renumbering so that each requirement (and all references to that requirement) are modified as requirements are moved, added, and deleted.

A simple Internet search reveals that there are approximately two dozen requirements-management tools commercially available (such as DOORS,

RML, RDD-10, and Requisite Pro, to name a few) [4]. Rather than attempt to review a subset of the tools, this section describes some of the most important features that a requirements-management tool should have to assist during the system-development process.

Requirements-management tools must support the initial capture of requirements, which may involve importing requirements statements from existing requirements documentation or extracting individual requirements from within a larger, more general source document. Once the requirements have been captured, the management tool must support the creation of requirements attributes. Attributes may be used to describe acceptable performance levels, requirement priority, testing approaches, and so on.

Requirements change over time. Performance levels may change, new requirements may be added, and old requirements may be deleted. The management tool must support this configuration-management function. Other configuration-management functions such as change proposals and approvals and change history should also be available.

Traceability has been emphasized throughout this text as being critical to the systems engineering process. Requirements-management tools must support and maintain traceability between the different levels of requirements (system to subsystem to component and so on) and between functional design and physical architecture. Closely associated with traceability is the ability to perform impact analysis prior to making changes. Most modern requirements-management tools support some form of impact analysis to enable uses to investigate the potential impact on the system of adding, deleting, or modifying requirements.

This text has described some of the attributes of "good" requirements, including completeness, clarity, consistency, and testability. Some requirements-management tools assist with the aim of producing good requirements by providing automated functions to check for inconsistencies and ambiguities. These functions are valuable, but users should initially establish a degree of confidence in the process and always cross check that the requirements are well written.

During preliminary design, a physical architecture is established and functional requirements allocated to it. Some requirements-management tools allow the physical architecture to be documented within the tool and support the process of requirements allocation. This feature, combined with an ability to perform impact analysis, becomes extremely valuable as the system development passes through the more detailed design stages.

The final functionality that all tools must possess is an ability to export the requirements in a usable form. This will require an ability to perform some formatting of the information, and most tools are delivered with built-in specification templates based on popular specification standards. The tools also normally allow the user to design their own specification format. Related to the requirement to export the requirements is an ability to interface with the most popular of office automation tools, such as spreadsheets, databases, and word processes, and to interface with other requirements-management tools. This ensures that the overall requirements-management process can be maintained even when customers, contractors and subcontractors make use of different tools.

7.1.5   Difficulties in Developing Requirements

Developing effective system-level requirements can be difficult and challenging. There is an ever-present urge to describe how the system should operate or be maintained. This urge must be overcome and requirements (at the system level) must be limited to describing what the system must do to satisfy the need. Clarification or justification, in the form of a rationale, should be provided in the form of why system-level requirements exist. This clarification and justification may assist in later processes by avoiding misinterpretations of requirements.

It is highly unlikely in a complex system development that one user (or a single group of users, for that matter) is able to articulate all system-level requirements. Additionally, the language employed by users in most complex systems tends to be arcane to outsiders and even to other specialist users of the same system. Some considerable time may have to be invested by requirements engineers to be able to comprehend the domain knowledge of all users.

Users sometimes identify particular systems or solutions that appear to meet their requirements. Instead of concentrating on clearly and completely defining the system-level requirements, users describe the functionality of the system they have seen in a marketing campaign. This is dangerous, as it can preclude potentially effective solutions from consideration. It can, however, be used to advantage in the requirements-engineering process because the user can at least visualize something that meets the perceived need.

Users in modern "right-sized" organizations are probably busier than they have ever been at any other point in time. The pressures of their daily workplace may preclude them from being sufficiently available to contribute to the requirements-engineering process. Change is also a constant factor in most users' work lives, and they may initially show some understandable animosity to the suggestion that they should devote valuable time to contributing to yet another change to the way in which they conduct business.

Organizations and systems involve people, and the human element is a strong influence on requirements engineering. Political, organizational, or personal bias is therefore often evident in early statements of requirement. Requirements engineering is not concerned with removing these factors, which will arguably exist whenever humans are involved in the process. Bias does, however, need to be identified as early as possible to ensure that subsequent decisions are made on an informed basis and the bias is formally recorded as a constraint.

Often poorly defined requirements are stated because the real requirement is not known or understood. It does not necessarily follow that the potential users of the system know what they want from the system. Requirements engineers must recognized this and try a number of different communications and elicitation techniques. As a corollary, users sometimes firmly believe that they know exactly what they want until they are presented with a system that has their specified functionality. To overcome this, users must be presented with prototypes, simulations, and models as soon as possible to ensure that there is a strong correlation between the users' perceived requirements and the appropriate functional statements encapsulated in the system specification.

In this text we suggest that every time a functional requirement is stated (no matter how loosely or incoherently it is stated), it is good practice to immediately articulate how well the function is to be achieved (performance requirement) and how the requirement is to be tested (verification requirement). This practice helps avoid incorporating requirements that reflect unrealistic user expectations.

Ambiguity is another result of poorly researched or understood requirements. Instead of specifying the requirement in precise terms, it is stated in an ambiguous manner. This is sometimes the result of an error on the author's behalf but often results from a poor understanding of the actual system-level requirement. Ambiguity is another difficulty in requirements definition that can be avoided by insisting that each requirement is testable. Good requirements are therefore clear, concise, testable, and do not contain ambiguous language.

Conflicting requirements often result when more than one person is responsible for drafting the system-level requirements. Conflicts occur when two or more requirements state mutually exclusive goals. Satisfaction of one requirement will result in another requirement failing to be satisfied. Conflicts must be resolved early in the systems engineering life cycle, as conflicts can be a source of ambiguity. Conflicts can also occur where ostensibly the same requirement appears twice in the same specification worded in slightly different ways. All requirements must be unique and stated only once to avoid confusion and conflict. It is often difficult to know where to place requirements in a framework such as the RBS because the requirement appears to be relevant to more than one section. For example, rate of climb in the aircraft system will invariably result from a regulatory constraint on the system. But rate of climb is also a performance requirement of the system. To that end, rate of climb could go in either the functional or the constraint section of the RBS presented in Figure 2.6. In general, a decision should be made and the requirement listed in one place only, which avoids ambiguity, enhances readability, and simplifies change management.

Users sometimes become confused between a need and a want. It is vitally important that system-level requirements reflect the true needs of the system and not be merely a wish list of additional functionality. This reinforces the need for forward and backward traceability between the endorsed SRD and the system specification. The additional functionality may add untold complexity and expense onto the system without enhancing the system's ability to meet the true system-level requirements. This gold plating of the requirements must be avoided.

The final challenge is to know when enough is enough, as it is difficult to judge when the requirements-analysis effort is complete. Although experienced personnel are best placed to make this judgment, no two system developments are likely to be the same, and the decision will require some consideration.

7.2   Synthesis—Various Tools

There are a number of tools available for use during the synthesis stage of the systems engineering process. This section aims to introduce the more common tools and briefly explain their characteristics.

7.2.1   Schematic Block Diagrams

Schematic block diagrams (or simply schematics) are perhaps the most basic and popular of the synthesis tools. Schematics allow designers to visually describe a solution to a given functional need as defined during the analysis stage of the process. Schematics show all of the elements of the solution and show their interrelationships and interfaces. These interfaces will include internal interfaces (between schematic elements) and external interfaces (with other subsystems). Schematics can be drawn at different levels of detail during systems engineering processes. For example, high-level schematics drafted during preliminary design will be far less detailed than the schematics used to assist with the detailed design and development of subsystem elements.

An example of an avionics system schematic block diagram is shown in Figure 7.3, showing the major subsystem elements, their interrelationships and internal interfaces, and external interfaces. Notice how easily the design and operation of this avionics subsystem can be understood.

A major advantage of schematic block diagrams is their ability to clearly communicate the proposed solution to technical and nontechnical persons alike. The phrase *a picture is worth a thousand words* adequately summarizes the advantages of the schematic.

7.2.2 Physical Modeling

Physical modeling can take any one of a number of forms. Perhaps the best known physical models are engineering models, such as the wind-tunnel models used during vehicle design to determine aerodynamic stability and characteristics. These models can include models of aircraft, land vehicles,

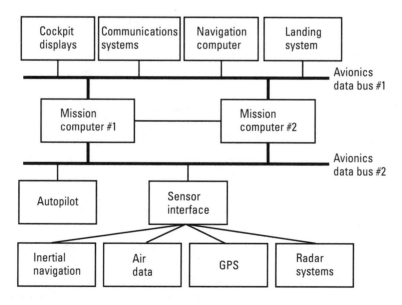

Figure 7.3 Sample schematic block diagram.

and sea-going vehicles such as ships and submarines. They are normally scale models and are built to exacting standards to ensure that the results obtained during the trials will be representative of the full-scale variant. Full-scale mockups of vehicles are also sometimes built to allow human-factors engineers to experiment with such design decisions as cable routing, access points, and operator equipment.

Electrical and electronic design can also benefit from physical modeling. Examples include *breadboards* of electrical and electronic circuits and *brassboards* used during the design of RF elements. These mockups are also called physical models.

It is fair to say that the use of physical models in today's design and development environment is gradually reducing as the power and capability of CAD tools increases. Wind-tunnel testing using scale models is very expensive and time consuming. CAD tools are now being used to at least supplement this form of modeling to help reduce the time and cost associated with development. Electrical and electronic design software is also improving and replacing the requirement to develop circuits using mockups. However sophisticated CAD tools become, physical modeling will still play a role in the synthesis of certain design solutions.

7.2.3 Mathematical Modeling and Simulation

Mathematical modeling and simulation are widely used and becoming increasingly popular as computers and software become more capable. As the name suggests, mathematical models rely on a mathematical representation of a physical system or design. The models can then be used to determine both static and dynamic characteristics of the system. An example of a static model would be one that represents a bridge's structural characteristics. Loads and stresses can be changed in magnitude and position to ensure that the structure will withstand the loads associated with its intended purpose. An example of a dynamic system is an aircraft's autopilot system. This is a classic example of a control system. Characteristics such as stability and response times can be determined using this sort of model.

Finite element analysis (FEA) [5] is a well-known mathematical modelling technique and serves as a good example of how mathematical modelling can assist and enhance synthesis throughout the systems engineering process. FEA is a mathematical technique used to study the static and dynamic behavior of physical structures such as aircraft wings when they are placed under varying conditions. FEA is used to understand this behavior without the need for expensive and time-consuming prototypes and tests. The structure under consideration is assumed to consist of a finite arrangement of individual elements. Each element is modelled mathematically using relatively straightforward physical and engineering laws. The individual models are then combined in such a way as to represent and model the entire structure. The model of the entire structure is analyzed to assist the iterative design (synthesis) of the structure.

FEA assists by ensuring that the design of the structure adequately meets the requirements assigned to it during the systems engineering process. It also ensures that the structure is not unnecessarily overengineered, saving valuable time and money during design.

Models and simulations are excellent tools for performing what-if analyses and determining the limits of the design. Variables can be changed and conditions easily altered to allow designers to investigate the likely behavior of the design in varying conditions. Simulators are widely used for a variety of different synthesis tasks. Environmental design and human-machine interface design are two examples of synthesis tasks that can benefit from simulation. The fidelity of simulators must be defined and confirmed prior to their use as synthesis tools.

To build and use a mathematical model, designers need to define the problem and formulate the model to replicate the problem. This model can be used with pen and paper if desired, but the more common approach is to convert the model into a computer program for ease of use and speed. Once this is done, perhaps the most crucial step must be performed. The performance of the model must be verified to ensure that the model is an accurate representation of the problem. It is pointless collecting data from a model if the model is incorrect (not representative). Once the model has been built and verified, designers can experiment with it and collect important design data.

7.3   Evaluation—Trade-off Analysis

Throughout the systems engineering management and processes chapters of this text, references are made to the evaluation of alternative solutions and the selection of the preferred approach. The process of analyzing alternatives and selecting the preferred solution is often generically called trade-off analysis. The analysis process must consider the different alternatives and balance the many (potentially conflicting) requirements prior to selecting the preferred solution. It must be emphasized that trade-off analysis is a decision-making aid only and should be used by the decision makers as a tool in making decisions. Trade-off analysis should not be considered the decision maker.

Trade-off analysis is used to assist decision making throughout all phases of the system life cycle. Table 7.1 summarizes typical activities conducted during the different life-cycle phases that may benefit from the use of trade-off analysis [6].

Analysis Process

The process employed when conducting a trade-off analysis will vary with the individual needs of the decision-making process. This section provides a guide to the general process that may be employed.

Definition of requirements.   The requirements and needs of the problem to be investigated must be documented in precise terms. Systems engineering documentation, including specifications and design aids such as function flow block diagrams, will provide a good source of this information. Sources will vary as the development moves through system life cycle.

Identification of alternative solutions.   The trade-off analysis must identify the widest range of alternative solutions possible to maximize the chance of selecting the optimal solution to the problem. There are many sources of alternative solutions, including systems engineering synthesis activities, COTS equipment, and other existing equipment. Once the range of solutions has

Table 7.1
Trade-off Analysis Examples

Acquisition Phase	Potential Applications
Conceptual design	Investigation of candidate new and emerging technology Selection of the preferred system design concept Identification of potential system configurations
Preliminary design	Selection of the technology to be employed Selection of the preferred system configuration
Detailed design	Consideration/selection of subsystem and component design Consideration/selection of testing methodology Optimize use of design space
Construction/ production	Consider design changes

been identified, the possible solutions should pass through a screening process to eliminate infeasible or unsuitable solutions from further consideration. This will avoid investing valuable time and effort on solutions that ultimately cannot be implemented for one reason or another. The remaining (feasible) solutions should be described in sufficient detail to allow further detailed consideration at a later stage in the process.

Nomination of selection criteria. Criteria must be formulated that provide a measure of the effectiveness of the alternatives to meet the requirements of the problem. If possible, these measures should be quantitative in nature to avoid subjective assessment of the alternatives as much as possible. The different criteria should be independent from one another to avoid assessing the same characteristics a number of times, and the criteria should be directly related to the defined problem.

Determination of criteria weighting. Once the criteria have been identified, they should be weighted according to their relative importance. Some criteria may be considered equally important as another and be assigned the same weighting; however, in most cases, there will be a clear difference in the level of importance among the criteria. This should be reflected in weighting. If considered necessary, the weightings may be withheld from the individuals performing the evaluation to help ensure the objectivity of the assessment. Numerical assessment through the use of weightings will assist in the comparative assessment of the alternatives under consideration. By combining an assessment of how well an alternative meets a certain criterion with a measure of the importance of that criterion, the evaluation will provide an excellent indication of the alternative's performance.

Example 7.1: Aircraft System Weighting

Figure 7.4 illustrates criteria weighting for our aircraft example.

Scoring function. It may be worthwhile to assign scoring functions to each criterion to ensure that the performance of each of the alternatives against the criteria is scored consistently. The scoring functions will be used by the scorers during the assessment of each alternative. A range of scores should be selected (for example from 0 to 10), and this range applied to all criteria. By applying different scoring ranges to different criteria, inadvertent weighting will be introduced. Weighting has already been assigned during the previous step.

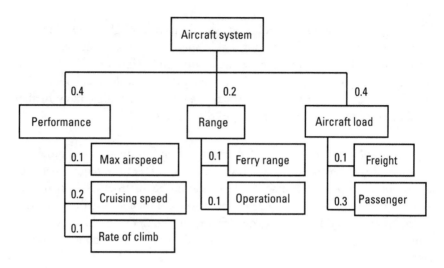

Figure 7.4 Aircraft system weighting.

Example 7.2: Aircraft System Scoring Function

Using the aircraft example, the scoring function in Figure 7.5 may be applicable to the cruising-speed criterion.

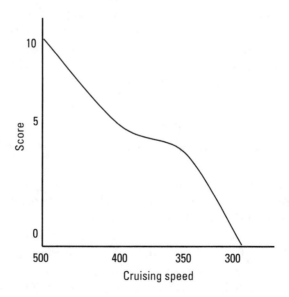

Figure 7.5 Aircraft system scoring function for the cruising-speed criterion.

In this example, alternatives that provide a cruising speed of greater or equal to 500 knots will score 10 points. Speeds of less than 300 knots will be given a score of 0.

Evaluation of alternatives. Once all the weights and scoring functions have been produced, the alternatives can be assessed. Normally, the scores and weightings will be recorded in a tabulated form and the total score for each alternative will be calculated using the following (or similar) expression

$$\text{Weighted score} = \sum_{i=1}^{N} w_i s_i$$

where

N = number of criteria
w_i = weighting of ith criteria
s_i = score of ith criteria

Provided the criteria, weightings, and scoring functions have been correctly compiled, the weighted scores will then provide an indication of the relative order of merit of each of the alternatives.

Sensitivity study. A sensitivity study, or *sensitivity analysis,* should be performed to confirm the validity of the results of the overall trade-off analysis. The sensitivity check will concentrate on the range of possible scores that could have been achieved by each alternative against each criterion (given that the score assessment may have been subjective to an extent). The range of scores will then translate into a range of weighted scores for each alternative. Sensitivity studies attempt to avoid the situation whereby the incorrect alternative is selected due to poorly informed assessment by investigating the situation where a small change in a single score can have significant effect on the overall weighted score. If the score ranges still clearly identify the preferred alternative, the sensitivity study will have reinforced the validity of that alternative. If, on the other hand, the score ranges overlap, the decision makers must consider the following courses of action:

- Obtaining additional information for each alternative to further differentiate between them;

- Reviewing the criteria, weightings, and scoring functions to ensure that they are correct;

- Ensuring that risk-management planning is in place to handle the possibilities arising from potential incorrect selection.

Endnotes

[1]   In addition to information contained in the general systems engineering references listed at the end of Chapter 1, specific detail on requirements engineering can be found in the following:

Andriole, S., *Managing Systems Requirements: Methods, Tools and Cases*, New York: McGraw-Hill, 1996.

Davis, A., *Software Requirements: Objects, Functions and States*, Engelwood Cliffs, NJ: Prentice Hall, 1993.

Gause, D., and G. Weinberg, *Exploring Requirements: Quality Before Design*, New York: Dorset House, 1989.

Grady, J., *Systems Requirements Analysis*, New York: McGraw-Hill, 1994.

Jackson, M., *Requirements and Specifications: A Lexicon of Software Practice, Principles and Prejudices*, Reading, MA: Addison-Wesley, 1995.

Kotonya, G., and I. Sommerville, *Requirements Engineering: Processes and Techniques*, West Sussex, England: John Wiley & Sons, 2000.

Leffingwell, D., and D. Widrig, *Managing Software Requirements: A Unified Approach*, Reading, MA: Addison-Wesley, 2000.

Macaulay, L., *Requirements Engineering*, London, England: Springer-Verlag, 1996.

Purdy, D., *A Guide to Writing Successful Engineering Specifications*, New York: McGraw-Hill, 1991.

Robertson, S., and J. Robertson, *Mastering the Requirements Process*, Harlow, England: Addison-Wesley, 1999.

Sommerville, I., and P. Sawyer, *Requirements Engineering*, Chichester, England: John Wiley & Sons, 1997.

Wieringa, R., *Requirements Engineering: Frameworks for Understanding*, New York: John Wiley & Sons, 1996.

Young, R., *Effective Requirements Practices*, Reading, MA: Addison-Wesley, 2001.

[2]   Specific detail on analysis tools can be found in the following:

Blanchard, B., and W. Fabrycky, *Systems Engineering and Analysis*, Upper Saddle River, NJ: Prentice Hall, 1998.

Coad, P., and E. Yourdon, *Object Oriented Analysis*, Upper Saddle River, NJ: Prentice Hall, 1990.

Defense Systems Management College, *Systems Engineering Management Guide*, Washington, D.C.: U.S. Government Printing Office, 1990.

Eisner, H., *Essentials of Project and Systems Engineering Management*, New York: John Wiley & Sons, 1997.

Gause, D., and G. Weinberg, *Exploring Requirements: Quality Before Design*, New York: Dorset House, 1989.

Goodland, M., and C. Slater, *SSADM: A Practical Approach*, London: McGraw-Hill, 1995.

Grady, J., *Systems Requirements Analysis*, New York: McGraw-Hill, 1994.

Heap, G., J. Stanway, and A. Windsor, *A Structured Approach to Systems Development*, London, England: McGraw-Hill, 1992.

Hitchins, D., *Putting Systems to Work*, Chichester, England: John Wiley & Sons, 1992.

Kotonya, G., and I. Sommerville, *Requirements Engineering: Processes and Techniques*, West Sussex, England: John Wiley & Sons, 2000.

Lacy, J., *Systems Engineering Management: Achieving Total Quality*, New York: McGraw-Hill, 1992.

Leffingwell, D., and D. Widrig, *Managing Software Requirements: A Unified Approach*, Reading, MA: Addison-Wesley, 2000.

Macaulay, L., *Requirements Engineering*, London, England: Springer Verlag, 1996.

Maier, M., and E. Rechtin, *The Art of Systems Architecting*, Boca Raton, FL: CRC Press, 2000.

Martin, J., *Principles of Object-oriented Analysis and Design*, Upper Saddle River, NJ: Prentice Hall, 1993.

Martin, J., *Systems Engineering Guidebook: A Process for Developing Systems and Products*, Boca Raton, FL: CRC Press, 1997.

Martin, J., and J. Odell, *Object-Oriented Analysis and Design*, Upper Saddle River, NJ: Prentice Hall, 1992.

Rumbaugh, J., et al., *Object-Oriented Modeling and Design*, Upper Saddle River, NJ: Prentice Hall, 1991.

Sage, A., and J. Armstrong, *Introduction to Systems Engineering*, New York: John Wiley & Sons, 2000.

Shlaer, S., and S. Mellor, *Object-oriented Systems Analysis*, Upper Saddle River, NJ: Prentice Hall, 1988.

Stevens, R., et al., *Systems Engineering: Coping with Complexity*, Hertfordshire, England: Prentice Hall, 1998.

Young, R., *Effective Requirements Practices*, Reading, MA: Addison-Wesley, 2001.

Westerman, H., *Systems Engineering Principles and Practice*, Norwood, MA: Artech House, 2001.

[3]   Based on the guidance contained in IEEE STD 1233-1998, *IEEE Guide for Developing System Requirements Specifications*, New York: IEEE, 1998.

[4]   See also Andriole, S., *Managing Systems Requirements: Methods, Tools and Cases*, New York: McGraw-Hill, 1996.

[5]   For more detail, see, for example, Cook, R., D. Malkus, and M. Plesha, *Concepts and Applications of Finite Element Analysis*, New York: John Wiley & Sons, 1989.

[6]   Defense Systems Management College, *Systems Engineering Management Guide*, Washington, D.C.: U.S. Government Printing Office, 1990.

8

Related Disciplines

8.1 Introduction

This chapter introduces some of the related disciplines that will interact with systems engineering efforts during the course of a normal project. Particular attention is given to the project-management discipline, as systems engineering is seen as being subordinate to project management in most projects.

This chapter is not intended to treat any of the related disciplines in a comprehensive manner. Each discipline is a large subject area in itself. Rather, the intention is to show how each of the major disciplines is related to systems engineering.

8.2 Project Management

Project management is an extremely large discipline that is documented comprehensively in a range of texts elsewhere. The detail is not replicated here, however. This section aims to introduce the discipline of project management and to detail the strong relationship between systems engineering and project management.

The strong relationship exists between systems engineering and project management primarily because every technical decision made as part of the systems engineering processes has an impact on the project management and vice versa. One way to introduce project management and investigate its relationship with systems engineering is to investigate the nine project management knowledge areas as defined in the PMBOK [1]. The PMBOK uses the same system life cycle described in this text. A summary of the following nine project management knowledge areas detailed in the PMBOK is provided in the following paragraphs.

Integration management ensures that all elements of the project are properly integrated and coordinated. The coordination involves making trade-offs between conflicting alternatives to ensure that the needs and expectations of the key project stakeholders are met. The key to integration management is the development of a project plan, which results from the entire project planning processes to form a governing approach to control and execution. The plan must be cognizant of the constraints and organizational policies as well as the outputs from the project planning processes. Once the plan is in place, the project-management function ensures that the plan is executed in accordance with its contents.

Scope management is primarily concerned with ensuring that the project includes all of the work required to successfully complete the project without including any additional and unnecessary work. The first stage in scope management is to develop a written scope document as the basis of the management effort. The PMBOK references the use of systems engineering techniques to assist with this task. Once the scope document has been written, the project must be broken into smaller, more manageable components. This breakdown is often performed using a WBS. The scope document and the WBS must then be formalized through an acceptance process to ensure that all parties are committed to the same scope. Once approved, scope management must take control of any changes to the approved project scope.

Time management is a key project-management activity that is sometimes referred to as schedule management. To manage the project's schedule, the project manager must start with a comprehensive list of the activities that make up the project. Tools such as the WBS will assist in this identification. These activities will normally be performed in a particular sequence that is often dictated by the systems engineering aspects of the project. For example, the activity "hardware/software integration" can only be performed following the activities "software test and acceptance" and "hardware test and acceptance." Other complications with respect to time management eventuate when activities have specific start and end dates that cannot be altered. Once the activities and their sequences have been determined, the durations for each of the activities will need to be estimated. Based on these pieces of information, the project manager can determine the schedule for the entire project. It then becomes an exercise in schedule control as activities are added, deleted, or changed throughout the project. There are many tools available to today's project manager to assist with the time-management function. These tools range from very capable automated software packages to reliance on expert judgment to predict activity duration.

Cost management is another major area of project management responsible for ensuring that the project is completed within the approved budget. The first task in determining the budget, called resource planning, is to determine the resources required to complete the project activities. Resources include personnel, materials, and equipment. Based on the resources required for each activity, the project manager can determine an estimate of the costs associated with each activity and the total cost of the entire project. As with the other project-management functions, changes in the project have an impact on the cost aspects of the project. The cost impact of these changes must be monitored and reported on an ongoing basis. Tools are available to assist with cost management, but a great deal of the estimation and control effort relies on expert judgement and experience.

Quality management is described in Section 8.3 as a discipline related to project management. The project manager must participate in the quality-assurance effort to ensure that quality is taken seriously by all involved. The first stage in the process is to plan the quality management effort by identifying which quality standards are to be applied to the project and how these will be satisfied. The overall project performance must be evaluated regularly to provide users with the confidence that the selected standards are being satisfied.

Human-resource management is not peculiar to project management but is, nonetheless, a vital part of the project-management function. The human-resource-management role ensures the most effective use of the personnel involved with the project. These personnel include not only those working on the project but other stakeholders, including customers and sponsors. The first job of the project manager is to acquire and organize the personnel required to perform the activities associated with the project. Once this has been done, the project manager must maintain the unity and direction of those involved with the project. This involves sound leadership and management skills, such as negotiation, communication, motivation, and team building.

Communications management is a fundamental requirement of any management situation, and project management is no exception to this rule. Communications is required to ensure that relevant project information is collected and disseminated to the appropriate personnel. The project manager must determine the information requirements of the key stakeholders in the project and the preferred format for the information to be provided to them. Once these decisions have been finalized, the project manager must ensure that the information is distributed as planned. A particularly important part of communications is the report generation associated with status reporting and phase or project closure. The role of communications management is worthy of emphasis here as a project-management function where the systems engineer can provide a huge amount of value. The systems engineer is, in some respects, part management staff and part technical staff. Generally, systems engineers are conversant with both technical and management languages and are able to act as very effective translators between the different management layers that exist within organizations. Systems engineers can also assist with horizontal communications between different design teams. They can assist with technical issues such as design space consideration, interfacing issues, and technical conflicts.

Risk management is covered in some detail in Chapter 5. The systems engineering effort focuses on the technical risks associated with the project, whereas the project manager broadens the coverage of risk to include such areas as schedule, cost, and the other elements of project management as defined in sources like the PMBOK. Systems engineering supplies input to the risk management process at the request of the project manager. Project risk management involves the same steps as technical risk management, namely risk identification, risk quantification, and risk-response development and control. The main source of information for the task of risk management is intuition and experience.

Procurement management is aimed at ensuring timely supply of goods and services from outside the immediate project environment in support of the project. The initial step is to plan what products and services are required to complete the project and when these will be needed. These details should then be documented detailing the requirements and potential suppliers. The project manager should seek quotes or offers from the suppliers and select the best of the offers. In some cases, the procurement will be a one-time purchase but often a standing offer or contract will result. The project management will be responsible for managing the resulting contracts. Once the requirement for the procurement no longer exists, the project manager is responsible for the closure of ant extant contracts with suppliers.

Systems Engineering and Project Management

The strong relationship between project management and systems engineering is demonstrated by considering the nine project management knowledge areas just summarized. Kauffman [2] investigates this relationship and demonstrates a large intersection between the two professions by referring to the

PMBOK, and IEEE 1220, and EIA/IS 632. A summary of Kauffman's findings is shown in Figure 8.1, which has been updated to include the current versions of both ANSI/EIA-632 and IEEE 1220 (1998).

Figure 8.1 shows a strong relationship between the systems engineering processes contained in IEEE 1220, EIA/IS 632, and ANSI/EIA-632, and the project management knowledge areas. The exceptions in IEEE 1220 and EIA/IS 632 are cost management, human-resource management, and procurement management. One could argue, however, that systems engineering management also contributes to these areas. Trade-off analysis and other technical

		PMBOK knowledge areas								
		Integration management	Scope management	Time management	Cost management	Quality management	Human-resource management	Communications management	Risk management	Procurement management
IEEE 1220	Requirements analysis		■							
	Requirements verification		■			■				
	Functional analysis		■							
	Functional verification					■				
	Synthesis		■							
	Design verification		■			■				
	Systems analysis	■							■	
	Control	■	■	■						
EIA/IS 632	Requirements analysis		■			■				
	Functional analysis/allocation		■			■			■	
	Synthesis		■			■				
	Synthesis analysis and control	■				■		■	■	■
ANSI/EIA 632	Acquisition and supply	■			■	■	■		■	■
	Technical management	■	■	■	■	■		■	■	
	System design	■	■			■		■	■	
	Product realization	■				■			■	■
	Technical evaluation	■	■	■	■	■	■	■	■	■

Figure 8.1 Relationship between project management and systems engineering.

decision-making tools provide an input into the cost-management function, especially if cost is a constraint that must be considered during the decision-making process. Systems integration and the control and management of interfaces will necessarily involve the systems engineer in significant amounts of human-resource management, as interface problems often arise from conflicts between sets of technical personnel. Chapter 2 discusses make-versus-buy decisions during the synthesis process. These decisions will sometimes directly involve the systems engineer in the procurement-management function.

Based on the findings of Kauffman as expanded to include ANSI/EIA-632, and the enhanced roles of the systems engineering in such areas as cost management, human-resource management, and procurement management, it is clear that there is a strong relationship between project management and systems engineering. It is clear from Figure 8.1 that as ANSI/EIA-632 becomes increasingly accepted and popular, the nexus between systems engineering and project management will almost certainly be strengthened even further.

Figure 8.1 illustrates that project management and systems engineering are inextricably entwined and can only be partially separated. For example, it is clear that systems engineering is responsible for requirements engineering and system architecting, yet project management is essential to ensure that all tasks are completed in a coordinated way. Project management is responsible for management of cost and schedule, although this can only be meaningful with input from systems engineering.

8.3 Quality Assurance

The term *quality* as used in the context of quality assurance means "the totality of feature and characteristics of a product or service that bears on its ability to satisfy stated or implied needs" [3].

With this definition in mind, quality assurance aims to ensure that the products and services required from a system development effort are delivered at the required quality level. This requires planned managerial effort to perform the following:

- Identify the quality assurance objectives and standards;
- Focus on preventative measures rather than cures;
- Collect relevant data and use the data to analyze results;
- Establish and monitor quality performance measures;

- Perform quality audits.

A major role of quality assurance is to ensure that appropriate plans and procedures are in place to perform the design and development tasks. These plans must be well documented, approved by the appropriate personnel, and maintained and updated when necessary. It is not, however, satisfactory to merely have the plans and procedures in place. The organization must operate in accordance with those plans and procedures. The quality assurance role will perform periodic audits to ensure that the systems engineering effort is progressing in accordance with the documented process. The quality-assurance representative requires evidence of the plans and procedures being followed.

8.4 Logistics Support

The main aim of logistics is to affect the design of the system to maximize supportability over the entire system life cycle. A definition of *logistics support* is "to assure the effective and economical support of a system throughout its programmed life cycle" [4]. The logisticians will also be interested and involved with maintaining and, where possible, improving system reliability throughout the system's life cycle. The FRACAS was introduced in Chapter 4 as a systems engineering management issue aimed at doing just that. Both logisticians and systems engineers must combine skills and expertise to make the FRACAS process as effective as possible.

To meet this definition, a number of factors need to be addressed throughout the acquisition phase of the project. These factors collectively comprise the logistics function and are discussed individually in the following sections.

Maintenance personnel. Maintenance personnel are required to perform to maintenance and support activities throughout the life cycle. These activities include fault finding and removal/installation of equipment. The required number of personnel and appropriate skill levels must be considered as part of the supportability issue.

Training and training support. The maintenance personnel and operators of the system will require training throughout the system life cycle. Analysis and determination of the training requirements should form part of the acquisition phase of the project. Training refers to the conduct and planning of the actual training programs, and support refers to the facilities and other equipment needed to support the training. For example, this logistics function may deem it necessary to develop and build a training mock-up as part of the system project.

Supply support. This involves the planning and procurement activities associated with the spares (repairable units and assemblies), nonrepairable components, consumable, and any other special supplies necessary to support the piece of prime equipment. These activities need to be initiated during the acquisition phase of the project as soon as equipment selection commences. Supply support extends to include provisioning for any warehousing requirements needed to accommodate the spares.

Support equipment. Complex systems such as aircraft avionics systems require a host of specialized test and support equipment to allow effective maintenance and support. The support equipment function of logistics is responsible for procuring the necessary equipment to support the scheduled and unscheduled maintenance activities.

Computer resources. These are required to support the system maintenance activities. Resouces include technical data, information systems, and database structures.

Packaging, handling, storage, and transportation (PHST). All aspects of the system (prime equipment, support equipment, personnel, and spares) may need special considerations when it comes to PHST. These considerations need to be analyzed and emphasized by logistics personnel. An example of a PHST issue is the movement of a completed system from the development to the operational environment.

Maintenance facilities. These include all facilities required to support scheduled and unscheduled maintenance activities at all levels of maintenance. As with any facility requirement, requirements need to be analyzed early and plans developed to ensure that maintenance facilities are in place in time for the arrival of the prime equipment. Facilities development can be an extremely complex, costly, and time-consuming task. In some circumstances, the facilities requirements are considered a separate project from the main system.

Technical data, information systems, and database structures. Modern support systems make extensive use of electronic systems to assist with the maintenance and support tasks. Technical data which results from the systems engineering effort will need to be arranged and stored in such a way as to maximize accessibility. Information systems and database structures are certain to be involved in this arrangement and storage. Examples of technical data include maintenance instructions, modification instructions, or supplier data.

Although the logistics functions have been dealt with independently in this section, it must be stressed that the elements are strongly interrelated. It is the role of the logistician to ensure that the elements are considered during the systems engineering processes and throughout the system life cycle.

8.5 Operations

Operations is the generic name sometimes given to the user representatives (the operators) participating in the system design and development effort. Operations personnel are vitally important to the project, and they play differing roles as the project progresses through the different acquisition phases. For example, in conceptual design and preliminary design, the operators are likely to be heavily involved in the correct and complete definition of the operational requirements and the interpretation of user requirements. This is vital because these interpretations will flow through the entire systems engineering effort and result in specific system performance. During the latter stages of the acquisition phase, the operators will be involved in various testing activities. Depending on the nature of the system under development, aspects of the system's operation (and therefore testing) will only be possible using appropriately trained personnel. Consider, for example, the testing of an attack helicopter's control system or the operation of an air defense radar system capable of tracking 150 targets simultaneously.

8.6 Design Support Network

The design support network (DSN) is a network of personnel that is established to assist the systems engineer with decision making. The DSN is a critically important resource for the systems engineering on a project, especially if the systems engineer is relatively inexperienced. The composition of the DSN is documented in the SEMP and normally consists of experienced systems engineers from different parts of the organization. The DSN is used for a range of reasons from informal discussions and advice to formal review and approval of major design decisions.

The DSN concept allows organizations to delegate relatively high levels of design responsibility to junior systems engineers without exposing that engineer and the organization to excessive levels of technical risk.

8.7 Software Engineering

Software engineering is defined as "the technological and managerial discipline concerned with systematic production and maintenance of software products that are developed and modified on time and within cost estimates" [5].

Software engineering has a set of processes, management requirements, tools, and related disciplines that are very similar to systems engineering.

Software products pass through a development process that is tied into the standard system life cycle presented in this text. This text detailed the processes leading up to the identification of CIs, and described HWCIs and CSCIs. Each CI is managed individually and follows an individual development cycle.

Once the CSCIs have been identified, the software engineering process works on these CIs to progressively break them into a manageable size. The first stage in this decomposition sees the CSCI broken into a number of computer software components (CSCs), which are in turn broken further into a number of computer software units (CSUs). CSUs are the lowest level of software decomposition.

CSUs are normally assigned to software teams who perform the development and coding of the units. When integration and testing occurs, it normally occurs from the bottom up, where the CSUs are individually tested against their individual requirements. The CSUs are then integrated together to form the CSCs, and these CSCs are also tested to ensure that the integration effort has been successful. In the same way, CSCs are integrated into CSCIs, and integration testing is performed again. Once the CSCIs have been tested, the system-level integration can occur as detailed in this text.

Software engineering has some unique challenges that must be addressed jointly by the software and systems engineering specialists. Software is inherently difficult to test due to the number of variables often associated with its successful operation. The measurement of software reliability is also difficult. Software maintenance and support is often overlooked, as many assume that software does not require maintenance and support once fielded. Software integration with host hardware brings about its own set of challenges, as does the issue of firmware as a specialized category of software. These challenges mean that traditional approaches to design, development, integration, testing, and support derived from hardware-dominated environments are not always satisfactory in a software environment. To that end, software engineering is recognized as a specialty engineering discipline in this text, and systems engineers should treat it accordingly.

8.8 Hardware Engineering

Hardware engineering processes proceed in a similar manner to those of software engineering following the selection of the HWCIs. The HWCIs are progressively broken into components and units in a similar way to the CSCs and CSUs.

Hardware units are designed and developed, and then tested via unit-level testing. Once successful, the units are integrated and retested as they form the hardware components. This process is repeated to form and test the HWCIs. Finally, the HWCIs and CSCIs are combined and system-level testing is conducted.

Hardware and software engineers must work closely together to ensure that interfacing issues are identified and addressed. Interfacing issues are sometimes addressed via an ICWG, which meets regularly during the system development process.

As with software engineering, documentation and management of the development process will be managed on an individual HWCI basis. This allows different approaches to be used for each HWCI. For example, HWCI A may make use of COTS equipment with minor modifications, while HWCI B may require a complete design and development effort.

Endnotes

[1]   Project Management Institute Standards Committee, *A Guide to the Project Management Body of Knowledge*, Upper Darby, PA: Project Management Institute, 1996.

[2]   Kauffman, D., "Project Management and Systems Engineering: Where the Professions Intersect-Generate Synergy Not Conflict," *Proceedings of the Eighth Annual International Symposium,* INCOSE, 1998.

[3]   Kerzner, H., *Project Management—A Systems Approach to Planning, Scheduling, and Controlling,* New York: Van Nostrand Reinhold, 1995, p. 1,041.

[4]   Blanchard, B., *Logistics Engineering and Management,* Upper Saddle River, NJ: Prentice Hall, 1992, p. 11.

[5]   Fairley, R., *Software Engineering Concepts,* New York: McGraw-Hill, 1985.

List of Acronyms

AT&E   acceptance test and evaluation

CAD   computer-aided design

CAE   computer-aided engineering

CAM   computer-aided manufacture

CASE   computer-aided software engineering

CCB   configuration control board

CCP   contract change proposal

CDR   conducting a design review

CI   configuration item

CM   configuration management

CMM   capability maturity model

CMMI   capability maturity model integration

CMP   configuration management plan

COD   concept of operations document

COTS   commercial-off-the-shelf

CPM   critical path method

CSCI   computer software configuration items

CSC   computer software component

CSU   computer software unit

DDP   design-dependent parameter

DFD   data flow diagram

DoD   Department of Defense

DSN   design support network

DT&E   developmental test and evaluation

ECP   engineering change proposal

EMC   electromagnetic compatibility

FAIT   fabrication, assembly, integration, and testing

FCAs   functional configuration audits

FEA   finite element analysis

FFBD   functional flow block diagram

FMECA   failure mode effects and criticality assessments

FQR   formal qualification review

FRACAS failure reporting, analysis, and corrective action system

GPS global positioning system

HMI human-machine interfaces

HWCI hardware configuration items

ICD interface-control document

ICWG interface-control working group

ICWG interface-control working groups

IEEE Institute of Electrical and Electronics Engineers

INCOSE International Council on Systems Engineering

IPPD integrated product and process development

LCC life-cycle cost

LSC logistics support concept

MOE measure of effectiveness

MOP measure of performance

MTP master test plan

NDI nondevelopmental items

OCD operational concept document or description

OFP operational flight program

OSA open system architectures

OT&E operational test and evaluation

PA   process area

PCA   physical configuration audit

PDR   preliminary design review

PHST   packaging, storage, and transportation

PMBOK   project management body of knowledge

RAAF   Royal Australian Air Force

RBS   requirements breakdown structure

RFT   request for tender

RMP   risk-management plan

SDD   system design document

SDR   system design review

SE   systems engineering

SECAM   systems engineering capability assessment model

SECM   systems engineering capability model

SE-CMM   systems engineering capability maturity model

SEDS   systems engineering detailed schedule

SEI   Software Engineering Institute

SEMP   systems engineering management plan

SEMS   systems engineering master schedule

SOW   statement of work

SRD stakeholder requirements document

SRR system requirements reviews

SS supplier sourcing

SSADM structured systems analysis and design

STD state transition diagram

SW software engineering

T&E test and evaluation

TEMP test and evaluation master plan

TPMs technical performance measures

TRAP technical reviews and audits plan

TRR test readiness review

URD user requirements document

WBS work breakdown structure

About the Authors

Ian Faulconbridge received his BE and MEngSc degrees in electrical engineering from the University of New South Wales Sydney, Australia, in 1990 and 1999, respectively, and an M.B.A. in project management from the University of Southern Queensland, Queensland, Australia, in 1996. Since 1990, he has held a number of systems engineering and project management positions in the fields of avionics, simulation, and communications systems. He has been lecturing with the School of Electrical Engineering of the University of New South Wales at the Australian Defence Force Academy since 1999, where he is currently a senior lecturer. His research and teaching interests include data compression and processing, avionics and navigation systems, radar systems, systems engineering practice, and project management. He is the author of a book on radar and radar electronic warfare and coauthor of a book on systems engineering.

Mike Ryan received his BE, MEngSc, and Ph.D. degrees in electrical engineering from the University of New South Wales in 1981, 1989, and 1996, respectively. Since 1981, he has held a number of positions in communications and systems engineering and in management and project management. Since 1998, he has been with the School of Electrical Engineering of the University of New South Wales at the Australian Defence Force Academy, where he is currently a senior lecturer. His research and teaching interests are in communications systems (network architectures, electromagnetics, radio wave propagation, mobile communications, and satellite communications), information systems architectures, systems engineering, project management, and technology management. He is the editor-in-chief of an international journal, author of a number of articles on communications and information systems and of a book on battlefield command systems, and coauthor of books on communications and information systems, tactical communications electronic warfare, tactical communications architectures, and systems engineering.

Index

Acceptance test and evaluation (AT&E)
  construction and production, 136
  defined, 128
  focus, 130
  regression testing, 130–31
  role and timing of, 129
  See also Test and evaluation (T&E)
Acquisition phase, 6–8
  conceptual design, 6–7, 29–65
  construction and production, 8
  detailed design and development, 7–8,
      97–118
  preliminary design, 7, 67–96
  See also System life cycle
Allocated baseline, 67
  defined, 94
  establishing, 96, 145
Allocation, 54–55
  matrix, 77, 78
  process of, 79
Analysis, 19–20
Analysis tools, 199–214
ANSI/EIA-549-1998, 188
ANSI/EIA-632, 179–87
  Annex D, 186
  Annex E, 187
  annexes, 186–87
  Annex F, 187

Annex G, 187
  bottom-up realization, 184
  building-block concept, 185
  concepts, 183–84
  defined, 179–80
  development, 179
  enabling products and examples, 184
  life-cycle activities, 186
  organization and approach, 180
  processes, 180–81
  processes illustration, 181
  requirements, 181–83
  summary, 187
Appendixes, 138–39, 172
Audits, 121–27
  assurance provided by, 122
  configuration, 150–51
  functional configuration, 125
  major, 125–26
  management, 126–27
  physical configuration, 125–26
  scheduling, 123

Backwards traceability, 10, 41

Capability maturity models (CMMs),
      189–91
  aim, 189
  extant, 190

Capability maturity models (continued)
   integration, 194–95
   SECAM, 191
   SECM, 191
   SE-CMM, 15, 190–94
   types of, 190
   use of, 190
Cause-and-effect analysis, 117
Change-control forms, 151
Change-management process, 148
CMM integration, 194–95
   adoption, 195
   common features, 195
   products, 194
   See also Capability maturity models
      (CMMs)
Commercial off-the-shelf (COTS)
      products, 60, 86–87
   defined, 86
   modified, 87, 102
   procurement of, 102
   use decision, 87
Communications management, 227–28
Components, 11
   computer software (CSC), 234, 235
   defined, 4
   risk, 139–40
Computer-aided design (CAD), 103
Computer-aided engineering (CAE), 103
Computer-aided manufacturing (CAM),
      104
Computer-aided software engineering
      (CASE) tools, 104
Computer software components (CSCs),
      234, 235
Concept demonstration, 43, 133
Concept of operations document (COD).
   See Stakeholder-requirements document
      (SRD)
Conceptual design, 6–7, 29–65
   aims, 29
   defined, 6
   feasibility analysis, 41–43
   introduction, 29
   product of, 29
   requirements analysis, 43–60

stakeholder requirements identification,
      30–41
system design review (SDR), 63–65
system-level synthesis, 60–63
tasks performed during, 30
T&E activities, 132–33
Configuration
   audits, 150–51
   control, 147–50
   identification, 145–46
Configuration control board (CCB), 147
Configuration items (CIs), 73, 144
   computer software (CSCI), 145
   configuration management, 75
   defined, 75
   engine example, 77–79
   hardware (HWCI), 145
   selection factors, 146
   variation of, 76
Configuration management (CM), 112,
      144–52
   aims, 144
   baseline establishment, 145
   documentation, 151–52
   effectiveness, 144
   functions, 145–51
Configuration-management plan (CMP),
      151
Constraints
   design, 38
   enterprise, 33–34
   external, 34
   project, 33–34
Construction and production, 108–11
   activity, 8
   defined, 109
   issues, 109
   systems engineering applied to
      planning, 110
   T&E activities, 136
Context diagram
   defined, 38
   illustrated, 39
Contract change proposals (CCPs), 151
Contractors, 4, 16
Cost management, 227
Critical design review (CDR), 106–8

aims, 106, 107–8
defined, 106
Critical path method (CPM) charts, 141–42
Cryptographic material, 118
Customer organization, 4

Design compatibility, 108
Design-dependent parameters (DDPs), 79–80
assigned during preliminary design, 80
defined, 79
Design reviews, 105–8
critical, 106–8
equipment/software, 105–6
Design space
example, 89–93
optimal use of, 88–93
trade-offs within, 93
Design support network (DSN), 233
Detailed design and development, 7–8, 97–118
aids, 103–4
construction and production, 108–11
evaluation of, 107–8
integration, 100–103
introduction, 97–98
operational use and system support, 112–17
phaseout and disposal, 117–18
process, 99–100
process illustration, 99
prototypes, 104–5
requirements, 98
reviews, 105–8
tasks, 97–98
T&E activities, 134–36
Developmental test and evaluation (DT&E), 132
in conceptual design, 132–33
construction and production, 136
defined, 128
in detailed design, 134
functions, 129
in preliminary design, 133
responsibility, 130
role and timing of, 129

system utilization, 137
testing levels, 130
See also Test and evaluation (T&E)
Development specifications, 153–54
approval of, 96
defined, 77, 93, 153
types of, 154
See also Specifications
Deviations, 149
Draft system specification, 55, 62

ECSS-E-10A, 188
EIA/IS-632, 171–72
additional information and requirements, 172
appendices, 172
general standard content, 171
notes section, 172
SEMP, 172
summary, 172
systems engineering process, 171
Electrical interfaces, 83
Electronic interfaces, 83
Engineering change proposals (ECPs), 151–52
content, 152
defined, 151
Engineering changes, 111
Enterprise constraints, 33–34
identifying, 33
types of, 34
Environmental interfaces, 84
Environmental testing, 135
Equipment/software design review, 105–6
Evaluation, 20–21
External constraints, 34

Failure mode effects and criticality assessments (FMECA), 117
Failure reporting, analysis, and corrective action system (FRACAS), 112, 114–17
defined, 114
establishment, 115
MIL-STD-2155(AS), 115, 116
process, 115, 116
process illustration, 116
steps, 115

Failure reporting, analysis, and corrective
  action system (continued)
  successful, 115
Fault-tree analysis, 117
Feasibility analysis, 41–43
  defined, 41
  example, 43
  process summary, 42
  steps, 42
  See also Conceptual design
Feedback loop, 55
Finite element analysis (FEA), 216–17
Formal qualification review (FQR), 124–25
  conducting, 125
  defined, 124–25
  See also Technical reviews
Forward traceability, 10, 41
Functional analysis, 51–55
Functional baseline
  defined, 29
  establishing, 145
  initial, establishment of, 65
Functional configuration audit (FCA), 125
Functional-flow block diagrams (FFBDs),
  47–48
  defined, 47–48
  describing aircraft refueling, 71
  describing aircraft turnaround, 70
  illustrated, 48
  as intuitive concept, 69
  use example, 53
Functional requirements
  analyzing, 47
  defining, 47–49
Functional testing, 135

Hardware engineering, 235
Human resources management, 227
Hydraulic/pneumatic interfaces, 83–84

IEEE 1220, 173–79
  additional material and requirements,
    179
  defined, 173
  design verification, 178
  detailed design, 176
  engineering plan content, 178–79

FAIT, 176
focus, 173
functional analysis, 177–78
functional verification, 174, 178
general standard content, 173–75
life-cycle model, 175–76
management layer, 178
physical verification, 174
preliminary design, 176
production and customer support, 176
requirements analysis, 177
requirements validation, 174, 177
scope, 173
subsystem definition, 175
summary, 179
synthesis, 178
synthesis function, 174
system development, 175
systems engineering process, 177–78
systems engineering process illustration,
    174
Integration, 100–103
  bottom-up process, 103
  CMM, 194–95
  management, 157–59, 226
Interface-control document (ICD), 153
  approval of, 96
  defined, 82, 153
  importance of, 158
  information, 83
  reviewed, 159
Interface-control working group (ICWG),
    85
Interface identification/design, 82–85
  example, 84–85
  illustrated, 82
  See also Preliminary design
Interfaces
  electrical, 83
  electronic, 83
  environmental, 84
  hydraulic/pneumatic, 83–84
  N2 diagram, 100, 101
  physical, 83
  software, 84
Interface testing, 135

Logistics
  aim, 231
  production plan, 111
Logistics support, 60, 231–33
  computer resources, 232
  maintenance facilities, 232
  maintenance personnel, 231
  packaging, handling, storage, and
      transportation (PHST), 232
  supply support, 232
  support equipment, 232
  technical data, information systems, and
      data structures, 232
  training/training support, 231–32

Master test plan (MTP). *See* Test and
      evaluation master plan (TEMP)
Material specification, 155
Mathematical modeling/simulation,
      216–17
  building/using, 217
  FEA, 216–17
  uses, 216
Measures of effectiveness (MOEs), 36
Measures of performance (MOPs), 36
MIL-HDBK-61A(SE), 188
MIL-HDBK-881, 189
MIL-STD-490A, 189
MIL-STD-499B, 164–70
  additional information and
      requirements, 168–70
  cost effectiveness and, 170
  functional tasks, 168–69
  general standard content, 164–65
  leverage options, 169
  life-cycle stages, 165
  notes section, 170
  requirements analysis, 166
  SEMP, 166–68
  summary, 170
  synthesis, 166
  systems engineering process, 166
  systems engineering process illustration,
      167
  technical reviews and, 170
  terms, 165
MIL-STD-961D, 189

MIL-STD-973, 188
MIL-STD-1521B, 188
Modifications
  aircraft example, 113–14
  effect on systems engineering impact,
      114
  impact, 113

N2 diagrams
  defined, 100
  example for aircraft system, 102
  functions, 100
  interfaces, 100, 101
  physical items, 100
  sample, 101
Need statement, 34–35

Objectives, 35
Operational concept document (OCD). *See*
      Stakeholder-requirements
      document (SRD)
Operational requirements, 49
Operational scenarios, 35
Operational test and evaluation (OT&E)
  aims, 131
  in conceptual design, 133
  construction and production, 136
  defined, 129
  in detailed design, 134–35
  focus, 131
  in preliminary design, 133–34
  role and timing of, 129
  system utilization, 137
  See also Test and evaluation (T&E)
Operations, 233
Overspecification, 156

Packaging, handling, storage, and
      transportation (PHST), 232
Performance requirements, 49–50
  allocating, 55
  examples, 49–50
  types of, 49
PERT, 141
Phaseout and disposal, 117–18
Physical and configurational testing, 135
Physical configuration audit (PCA),
      125–26

Physical interfaces, 83
Physical modeling, 215–16
    forms, 215
    uses, 216
Preliminary design, 7, 67–96
    activities, 68
    evaluation of, 95
    interface identification and design,
        82–85
    introduction, 67–68
    RBS vs. WBS, 80–81
    requirements allocation, 72–80
    responsibility, 67
    review, 94–96
    subsystem-level synthesis and
        evaluation, 85–94
    subsystem-requirements analysis, 68–72
    T&E activities, 133–34
Preliminary design review (PDR), 68,
        94–96
    activities, 94
    aims, 94, 95–96
    approaches, 95
    conducting, 94
    major task, 95
    successful, 95
    See also Preliminary design
Process specification, 154–55
Procurement management, 228
Product baseline, 97
    defined, 98
    establishing, 108, 145
Production, 108–11
    engineering considerations, 110
    issues, 109
    plan, 110
    requirements, 109
    systems engineering applied to
        planning, 110
    See also Construction and production
Product specification, 154
    content, 154
    defined, 93, 154
    See also Specifications
Program summary, 138
Project constraints, 33–34
    identifying, 33

    types of, 34
Project management, 225–30
    defined, 225
    knowledge areas, 226–28
    systems engineering and, 228–30
Project Management Body of Knowledge
        (PMBOK), 21
Prototypes
    development, 104–5
    example, 105

Quality assurance, 230–31
    defined, 230
    role, 231
Quality factor testing, 135
Quality management, 227

Radioactive material, 117–18
Regression testing, 130–31
Request for tender (RFT), 61
Requirements
    analysis and negotiation, 208
    associated information, 205–8
    automated management tools, 210–12
    changes, 211
    conflicting, 213–14
    creep, 41
    criticality, 206
    defined, 200
    derived, 51
    detailed design, 98
    development difficulties, 212–14
    documentation, 209–10
    elicitation and generation, 201–8
    feasibility, 206–7
    functional, 47–49
    history, 207
    identification, 202–4
    management/maintenance, 208–9
    operational, 49
    performance, 49–50
    poorly defined, 213
    priority, 205–6
    production, 109
    properties, 204–5
    rationale, 50–51, 207
    relationship to other requirements, 207
    risk, 207

source, 207
system engineering, 59–60
system-level, 51–52
traceability, 10
type, 207
unendorsed, 41
unique identifier, 205
validation, 209
verification, 50
writing guidelines, 209–10
Requirements allocation, 72–80
defined, 72
matrix, 77, 78
process, 79
process illustration, 73
See also Preliminary design
Requirements analysis, 43–60
aim, 43
defined, 43
draft system specification, 55
focus, 44
framework establishment, 45–47
functional requirements definition,
47–49
performance requirements definition,
49–50
process summary, 44
rationale assignment, 50–55
SRRs, 58–59
system-level considerations, 59–60
TPMs definition, 55–58
verification requirements definition, 50
See also Conceptual design
Requirements breakdown structure (RBS),
45
defined, 45
example, 46
example after grouping and allocation,
54
populating, 55
system requirements in, 45
WBS vs., 80–81
Requirements engineering, 10, 200–209
defined, 10, 199
disciplined approach to, 18
iterative activities, 201
philosophy, 200

problems, 202
process, 200–201
Resources, 110
Risk identification, 140–41
defined, 140
risk sources and, 141
See also Technical risk management
Risk-response development and control,
142–43
Risk(s)
acceptance, 143
avoidance, 143
components, 139–40
management, 143, 228
management documentations, 143–44
mitigation, 143
quantification, 141–42
reduction, 17–18
See also Technical risk management
Risk-severity table, 142

Schematic block diagrams, 214–15
benefits, 214
sample illustration, 215
Scope management, 226
Scoping system, 38–40
context diagram, 38
defined, 38
example, 38–40
Scoring function, 219–21
Sensitivity study, 221–22
Software
engineering, 234
interfaces, 84
maturity determination, 108
Specifications, 152–55
defined, 93
development, 153–54
material, 155
process, 154–55
product, 154
standards, 189
system, 153
tree, 155
Stakeholder-requirements document
(SRD), 30–32
content example, 31–32

Stakeholder-requirements document
    (continued)
    defined, 30
    endorsement, 40–41
    populating, 40
    potential audience, 32
    reader, 31
    sections, 37
    structure confirmation, 36–37
    system specification and, 62
    system-specification language vs., 62–63
    traceability, 41
    writing of, 31
Stakeholder requirements identification,
        30–41
    external constraints, 34
    life-cycle concepts definition, 36
    MOEs definition, 36
    needs, goals, objectives definition,
        34–35
    operational scenarios definition, 35–36
    process, 32–33
    process summary, 33
    project and enterprise constraints,
        33–34
    scoping system, 38–40
    SRD, 30–32
    SRD endorsement, 40–41
    SRD population, 40
    SRD structure confirmation, 36–37
    traceability, 41
    See also Conceptual design
Stakeholders, 4
Standards, 156–57, 188
    overspecification, 156
    specification, 189
    systems engineering, 163–64, 188
    tailoring, 156–57
    unintentional invocation of, 157
Statement of work (SOW), 59
Status accounting, 150
Subcontractors, 4
Subsystem-level synthesis, 85–94
    alternatives investigation, 86–88
    design space use, 88–93
    illustrated, 85
    preferred solution selection, 93–94

subsystem design sources review, 85–86
    See also Preliminary design
Subsystem-requirements analysis, 68–72
    activities, 69
    example, 70–72
    See also Preliminary design
Subsystems
    aircraft example, 74
    avionics, 4, 74–75
    CIs, 75–79
    defined, 4
    engine, 4
    interior, 75
    optimality, 88
    road map, 88
Synthesis, 20
Synthesis tools, 214–17
    mathematical modeling and simulation,
        216–17
    physical modeling, 215–16
    schematic block diagrams, 214–15
System design document (SDD). See
    Stakeholder-requirements document (SRD)
System design review (SDR), 63–65
    functions, 64–65
    goals, 63
    process summary, 64
    See also Conceptual design
System-level synthesis, 60–63
    defined, 60
    process summary, 61
    See also Conceptual design
System life cycle, 5–8
    acquisition phase, 6–8
    defined, 5
    focus, 6, 11–13
    illustrated, 5
    life-cycle cost (LCC), 6
    money savings, 16–17
    phases, 5
    significance, 5–6
    systems engineering impact on, 17, 18
    utilization phase, 8
System requirements reviews (SRRs),
        58–59
    aim, 59
    number of, 58

Systems
   aircraft, 4
   defined, 2
   description, 138
   failures, 17
   in functional terms, 2–3
   hierarchical nature of, 4
   optimization and balance, 13–14
   in physical terms, 3
   prototype development, 104–5
   resources, 3–4
   scoping, 38–40
Systems engineering
   analysis, 19–20
   benefits, 16–18
   defined, 9
   development, 1
   emergence, 1
   evaluation, 20–21
   focus on life cycle, 11–13
   integration of disciplines and specialties,
      14
   introduction to, 1–24
   optimization and balance, 13–14
   processes, 23
   process tools, 199–222
   in project management, 228–30
   relevance, 15–16
   requirements, 10, 59–60
   standards, 163–64
   synthesis, 20
   system life cycle impact, 17, 18
   tools, 23–24
   top-down approach, 10–11
Systems Engineering Capability Maturity
      Model (SE-CMM), 15, 190–94
   capability levels, 192–94
   defined, 191
   focus, 191
   foundation, 192
   process areas, 193
   summary, 194
   See also Capability maturity models
      (CMMs)
Systems engineering framework, 21–24
   aim, 21
   defined, 21

   illustrated, 22
   related disciplines, 24
   structure, 22
   systems engineering management, 23
   systems engineering processes, 23
   systems engineering tools, 23–24
Systems engineering management, 14–15,
      23, 121–60
   configuration management, 144–52
   integration management, 157–59
   introduction, 121
   planning, 159–60
   specifications and standards, 152–57
   technical reviews and audits, 121–27
   technical risk management, 139–44
   test and evaluation (T&E), 127–39
   tools, 163–95
Systems engineering management plan
      (SEMP), 159–60
   approval process, 159
   defined, 159
   MIL-STD-499B, 166–68
   suggested contents, 159
System specification, 153
   approval of, 65
   defined, 153
   forms, 63
   reviewing, 72
   See also Specifications

Tailoring, 156–57
Technical demonstration, 43, 133
Technical reviews, 121–27
   assurance provided by, 122
   formal qualification (FQR), 124–25
   major, 123–25
   management, 126–27
   number of, 122
   scheduling, 123
   scope, 122
   test readiness (TRR), 123–24
   types of, 122
   See also Systems engineering
      management
Technical reviews and audits plan (TRAP),
      127
Technical risk management, 139–44

Technical risk management (continued)
  as ongoing concern, 140
  risk identification, 140–41
  risk-management documentation,
      143–44
  risk quantification, 141–42
  risk-response development and control,
      142–43
  as team effort, 139
Test and evaluation master plan (TEMP),
      124, 128, 137–39
  appendixes, 138–39
  contents, 138–39
  as contractual deliverable, 132
  defined, 137
  drafting, 132, 137–38
  outline, 138
  program summary, 138
  resource summary, 138
  system description, 138
  T&E outline, 138
  T&E resource summary, 138
Test and evaluation (T&E), 127–39
  acceptance (AT&E), 128, 130–31
  activities, 132–37
  conceptual design activities, 132–33
  construction and production activities,
      136
  defined, 127
  detailed design and development
      activities, 134–36
  developmental (DT&E), 128, 129–30
  operational (OT&E), 129, 131
  preliminary design activities, 133–34
  system utilization activities, 136–37
  test management, 132
Testing
  environmental, 135
  functional, 135
  interface, 135
  physical and configurational, 135
  quality factor, 135
  regression, 130–31
Test readiness review (TRR), 123–24
  defined, 123–24
  documentation reviewed, 124
  See also Technical reviews

Time management, 226–27
Top-down approach, 10–11, 12
TPMs
  defined, 55
  defining, 55–58
  example, 56–58
  identifying, 55
  investigation of, 106
  list of, 55, 56
  rate-of-climb, 57, 58
  reviewing, 72
  role of, 56
  tracking, 58
Traceability, 41
  assessment of, 96
  backwards, 10, 41
  forward, 10, 41
  matrix, 77
  support for, 10
Trade-off analysis, 60, 217–22
  as decision-making aid, 217
  definition of requirements, 218
  determination of criteria weighting, 219
  examples, 218
  identification of alternative solutions,
      218–19
  nomination of selection criteria, 219
  process, 217, 218
  scoring function, 219–21
  sensitivity study, 221–22
Type A specification. See System
      specification
Type B specification. See Development
      specifications
Type C specification. See Product
      specification

Use cases, 35
User requirements document (URD). See
      Stakeholder-requirements
      document (SRD)
Utilization phase, 8, 112
  activities, 8
  T&E activities, 136–37

Verification requirements, 50
V&V
  defined, 128

objective, 128

Waivers, 149–50

Work breakdown structure (WBS), 45
aircraft system, 81
concept, 80
RBS vs., 80–81

Recent Titles in the Artech House Technology Management and Professional Development Library

Bruce Elbert, Series Editor

Critical Chain Project Management, Lawrence P. Leach

Decision Making for Technology Executives: Using Multiple Perspectives to Improve Performance, Harold A. Linstone

Designing the Networked Enterprise, Igor Hawryszkiewycz

Engineering and Technology Management Tools and Applications, B. S. Dhillon

The Entrepreneurial Engineer: Starting Your Own High-Tech Company, R. Wayne Fields

Evaluation of R&D Processes: Effectiveness Through Measurements, Lynn W. Ellis

From Engineer to Manager: Mastering the Transition, B. Michael Aucoin

Introduction to Information-Based High-Tech Services, Eric Viardot

Introduction to Innovation and Technology Transfer, Ian Cooke and Paul Mayes

Managing Complex Technical Projects: A Systems Engineering Approach, R. Ian Faulconbridge and Michael J. Ryan

Managing Engineers and Technical Employees: How to Attract, Motivate, and Retain Excellent People, Douglas M. Soat

Managing Successful High-Tech Product Introduction, Brian P. Senese

Managing Virtual Teams: Practical Techniques for High-Technology Project Managers, Martha Haywood

Mastering Technical Sales: The Sales Engineer's Handbook, John Care and Aron Bohlig

The New High-Tech Manager: Six Rules for Success in Changing Times, Kenneth Durham and Bruce Kennedy

Planning and Design for High-Tech Web-Based Training, David E. Stone and Constance L. Koskinen

Preparing and Delivering Effective Technical Presentations, Second Edition, David Adamy

Reengineering Yourself and Your Company: From Engineer to Manager to Leader, Howard Eisner

Successful Marketing Strategy for High-Tech Firms, Second Edition, Eric Viardot

Successful Proposal Strategies for Small Businesses: Using Knowledge Management to Win Government, Private Sector, and International Contracts, Third Edition, Robert S. Frey

Systems Engineering Principles and Practice, H. Robert Westerman

Team Development for High-Tech Project Managers, James Williams

For further information on these and other Artech House titles, including previously considered out-of-print books now available through our In-Print-Forever® (IPF®) program, contact:

Artech House	Artech House
685 Canton Street	46 Gillingham Street
Norwood, MA 02062	London SW1V 1AH UK
Phone: 781-769-9750	Phone: +44 (0)20 7596-8750
Fax: 781-769-6334	Fax: +44 (0)20 7630-0166
e-mail: artech@artechhouse.com	e-mail: artech-uk@artechhouse.com

Find us on the World Wide Web at:
www.artechhouse.com